ANIMAL BEHAVIOR
An Evolutionary Approach

ANIMAL BEHAVIOR

An Evolutionary Approach

Victor S. Lamoureux

Adjunct Professor of Biology, Broome Community College,
Binghamton, New York, U.S.A.

Apple Academic Press

TORONTO   NEW JERSEY

Animal Behavior: An Evolutionary Approach

© Copyright 2011*
Apple Academic Press Inc.

First Published in the Canada, 2011
Apple Academic Press Inc.
3333 Mistwell Crescent
Oakville, ON L6L 0A2
Tel. : (888) 241-2035
Fax: (866) 222-9549
E-mail: info@appleacademicpress.com
www.appleacademicpress.com

The full-color tables, figures, diagrams, and images in this book may be viewed at www.appleacademicpress.com

First issued in paperback 2021

ISBN 13: 978-1-77463-245-1 (pbk)
ISBN 13: 978-1-926692-78-4 (hbk)

Victor S. Lamoureux

Cover Design: Psqua

Library and Archives Canada Cataloguing in Publication Data
CIP Data on file with the Library and Archives Canada

CONTENTS

INTRODUCTION

Human beings have been students of animal behavior from their earliest days. Understanding animals and what they did meant survival for humans, either through use of animals as food or avoidance of them as potential hazards. Certainly the understanding of animal behaviors and habits gave a selective edge to early humans who practiced the study. Further evidence for this appreciation of animal behavior can be seen in the many cave paintings depicting animals, while not depicting any of the other facets of the painters' lives. However, the modern study of animal behavior is rather recent and was first widely acknowledged with the awarding of the Nobel Prize in 1973 to Niko Tinbergen, Karl von Frisch, and Konrad Lorenz.

Animal behavior is technically known as ethology, which is considered the systematic study of the behavior of animals under natural conditions. Although historic distinctions existed between the fields of ethology and animal psychology, with the former being largely performed in natural settings and the latter in controlled laboratory settings, modern studies are frequently a combination of laboratory and field work. Ethology is highly integrative and draws on the fields of evolution, ecology, psychology, molecular biology, development, neurobiology, endocrinology, and mathematics. A student in an animal behavior course will typically already have a substantial background in math and sciences.

Tinbergen set the stage for the questions of animal behavior in his 1963 paper, "On Aims and Methods of Ethology". Now commonly referred to as "Tinbergen's Four Questions", he suggested four ways to answer the question "Why?" when referring to animal behavior: as a function of causation, evolution, function, and ontogeny. One aspect is no more important than another, but all four are needed to gather a clear picture of the role of a behavior. In most cases, behaviors should be considered adaptations and therefore subject to the pressures of natural selection.

The study of animal behavior has become a foundation to other biological disciplines, including neuroscience, behavioral genetics, ecology, and conservation biology. Although the study of animal behavior might seem only academic, or even esoteric, the reality is that ethology is applicable to many aspects of our modern lives. The comparative approach of studying animals and using them as models for ourselves leads to important discoveries on how the human brain and behaviors might work. Animal behavior studies are also useful to better the welfare of animals in a wide variety of situations, from agriculture to zoos.

One of the more recent areas in which animal behavior's importance has emerged is in the field of conservation biology. In fact, insights from animal behavior have become critical in the implementation of conservation strategies. The design of wildlife reserves and biological corridors linking reserves needs to be examined with a full knowledge of the wildlife utilizing such reserves. Animal behavior is also critical to the success of captive breeding programs and the subsequent release of these captive-reared individuals. Not fully understanding the behavioral repertoire of an endangered species will likely lead to ineffective conservation measures.

Hard work, long hours, and stiff competition await those who want to enter this field. Those with bachelor's degrees will have opportunities in research, pharmaceutical testing, animal training, and conservation. Specialization in animal behavior usually occurs at the graduate level with a PhD, or even as a veterinarian. Most positions will be in research with the government or research and teaching at colleges and universities. But for those who love animals and who want to delve into and discover new aspects of their behavior, the rewards are substantial. With the current worldwide decline in natural systems, it seems likely that a demand for people trained in animal behavior and conservation will continue into the future.

— Victor S. Lamoureux, PhD

ACKNOWLEDGMENTS AND HOW TO CITE

The chapters in this book were previously published in various places and in various formats. By bringing these chapters together in one place, we offer the reader a comprehensive perspective on recent investigations into this important field.

We wish to thank the authors who made their research available for this book, whether by granting permission individually or by releasing their research as open source articles or under a license that permits free use provided that attribution is made. When citing information contained within this book, please do the authors the courtesy of attributing them by name, referring back to theiroriginal articles, using the citations provided at the end of each chapter.

Free-Ranging Macaque Mothers Exaggerate Tool-Using Behavior when Observed by Offspring

Nobuo Masataka, Hiroki Koda, Nontakorn Urasopon
and Kunio Watanabe

ABSTRACT

The population-level use of tools has been reported in various animals. Nonetheless, how tool use might spread throughout a population is still an open question. In order to answer that, we observed the behavior of inserting human hair or human-hair-like material between their teeth as if they were using dental floss in a group of long-tailed macaques (Macaca fascicularis) in Thailand. The observation was undertaken by video-recording the tool-use of 7 adult females who were rearing 1-year-old infants, using the focal-animal-sampling method. When the data recorded were analyzed separately according to the presence/absence of the infant of the target animal in the target animal's proximity, the pattern of the tool-using action of long-tailed adult female macaques under our observation changed in the presence of the infant as compared with that in the absence of the infant so that the stream of tool-using action was punctuated by more pauses, repeated more often, and performed for a longer period during each bout in the presence of the infant. We interpret this as evidence for the possibility that they exaggerate their action in tool-using so as to facilitate the learning of the action by their own infants.

Introduction

The population-level use of tools has been reported in various animals. One of the best known instances of this is the so-called "ant-fishing" by free-ranging chimpanzees (Pan troglodytes) [1]. Nonetheless, how tool use, including that of ant-fishing in chimpanzees, might spread throughout a population is still an open question [2]. There is some controversy as to whether the transfer of these cultural practices is accomplished across individuals by observational social learning or just by individual learning alone [3].

Although there is some disagreement about whether or not various forms of observational social learning play a role in the transmission, there is a general consensus among researchers that the recipient is solely responsible for the successful acquisition of the skill, and that the skill's donor does not have any active role in the transmission of cultural information. In the present paper, on the other hand, we present evidence which indicates the possibility that free-ranging adult long-tailed macaques (Macaca fascicularis) modify their action in tool-using so as to facilitate the learning of the action by their own infants. The behavior we observed was that of inserting human hair or human-hair-like material between their teeth as if they were using dental floss. We compared the pattern of the behavior in each of 7 adult females when her own infant was in her proximity and when any other group member was not in her proximity.

Our study of the tool-using behavior in a group of the macaques in Thailand started in 2004 and continues up to the present [4]. Whenever the material picked up by an animal is to be used as the tool, the animal subsequently grasps the hair taut between its two hands. Then, the animal inserts the taut hair between its open jaws, and the action ends when the animal closes its jaws to engage the taut hair, and pulls the hair sharply to one side by one hand and removes it from its mouth. Here a 'bout' of the tool-use is defined as starting at the moment of grasping the material with the hands and ending at the moment of completely removing it from the mouth. With this removing action, food, if present could be cleaned from between the teeth. Before removing the hair, the animal was often observed to repeatedly rapidly close and open ("snap") its jaw to engage (clamp) the taut hair between its teeth. When this occurred, the number of times the animal clamped on the hair could be counted, calling it the number of snaps. Subsequent to the occurrence of such snapping, moreover, the animal was often observed to remove the taut hair which was kept grasped between the two hands, to briefly look at it at about eye level, and to reinsert it in its mouth as before. When this was observed, it was defined as an occurrence of "reinsertion" in a given bout. "Reinsertion" might be repeated in that bout: after reinserting the hair, the animal might repeat the same action and take out the hair again while grasping it with two hands. That bout continued until the animal finally pulled out the reinserted hair to one side using one hand. In each such bout, the number of occurrences of reinsertion as well as the number of occurrences of snapping while the hair was inserted could be counted. The length of each bout could also be measured by counting the number of frames of the video which were required to record from the onset until the end of the bout. In addition, the number of occurrences of "removing of the hair from the mouth" was computed in each bout as attempts to clean the teeth. It could be counted as 'X+1 (X = 0, 1, 2,,,)' in a given bout when the number of occurrences of reinsertion was 'X' in the bout.

When a bout ended, perhaps on the completion of the cleaning of the teeth, the animal abandoned the material onto the ground on some occasions. If this was observed, the tool-use 'episode' ended, during which a single bout of the activity was undertaken. Alternatively, however, the animal again grasped the material with the two hands and began another bout with an interval of no more than 1 second or so. Then, that episode continued until the animal finally abandoned the stimulus. Thus, the number of 'bouts' in the episode could be counted. Also, the number of frames of the video which were required to film from the onset of the first bout until the end of the final bout was defined as the total duration of that episode. If only a single bout was included in a given episode, the duration of the episode coincided with the duration of the bout. In addition, the total number of occurrences of "removing the hair from the mouth" in the episode was computed as an index of the frequency of cleaning attempts in the episode.

Results

Results of the analyses are summarized in Figure 1. When the average number of occurrences of reinsertion in a given bout of the tool-use was computed across subjects, a likelihood-ratio test revealed that the score when the infant was in the proximity of the target mother was greater than that when the infant was absent

($\chi 12$ = 22.201, p<0.0001). Similarly, the average number of jaw snaps during each insertion of the stimulus was greater when the infant was present as compared to when the infant was absent ($\chi 12$ = 123.6, p<0.0001). The average duration of a given bout when the infant was present was longer that that when the infant was absent ($\chi 12$ = 44.51, p<0.0001). In a given bout, the number of occurrences of reinsertion was found to positively correlate with the number of jaw snaps during each insertion (Pearson's correlation = 0.232, n = 355, p<0.01). In a given bout, both the number of occurrences of reinsertion and the number of jaw snaps were found to positively correlate with the duration of the bout (Pearson's correlation = 0.770, p<0.001; 0.417, p<0.001; n = 355, respectively). The average duration of a given episode, on the other hand, did not differ significantly when the infant was present compared to when it was absent (χ^1_2= 1.592, p = 0.2071) because the average number of bouts in a given episode when the infant was absent was greater than that when the infant was present (χ^1_2= 8.9008, p = 0.00285). The average number of occurrences of removal of the hair from the teeth in a given episode did not differ when the infant was present compared to when it was absent (χ^1_2 = 1.0519, p = 0.3051, mean±95%CI = 3.38±0.40 when the infant was present, and 3.76±0.50 when the infant was absent).

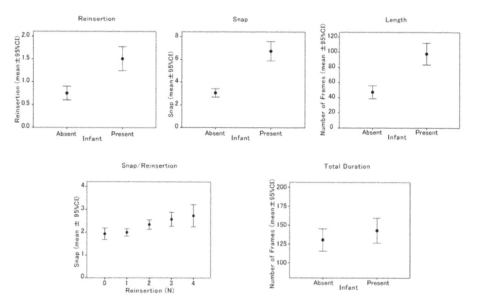

Figure 1. Summary of results of the analyses. Average scores of number of occurrences of reinsertion in a given tool-using bout (Reinsertion), of number of occurrences of snapping during each insertion (Snap), of length of each bout (Length), of overall mean number of snaps during each insertion as a function of number of occurrences of reinsertion in a tool-using bout (Snap/Reinsertion), and of total duration of a given tool-using episode (Total Duration) are computed across target adult females when the infant was in her proximity and when the infant was absent.

Discussion

Overall, once the long-tailed macaque mothers (the target animals) started to use the stimulus as a tool, they devoted a similar amount of time to the stimulus regardless of whether or not their infant was present. However, as shown in Figure 2, the pattern of their action changed in the presence of the infants as compared with that in the absence of the infants so that the stream of tool-using action was punctuated by more pauses, repeated more often, and performed for a longer period during each bout in the presence of the infants.

With Infant

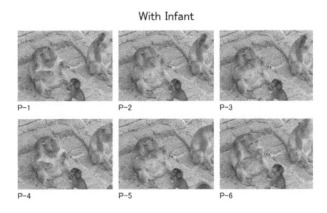

Without Infant

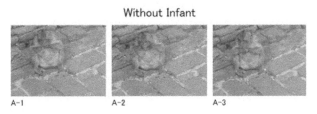

Figure 2. Typical sequences of the action of "flossing teeth." (P 1 to 6) When her infant was in the proximity of an adult female (With Infant; P-1: Grasp the hair taut, P-2: Insert, P-3: Snap, P-4: Look at the hair, P-5: Reinsert, P-6: Pull out). (A 1 to 3) When no animal was in the proximity of an adult female (Without Infant; A-1 Grasp the hair taut, A-2: Insert, A-3: Pull out).

As a possible factor affecting this difference, the activity of feeding by the animals per se is not considered likely because the present observations were undertaken at least 30 min after the end of the animal's final food-taking. Rather, it seems more likely that the behavioral difference is socially modulated, and influenced by the presence/absence of other animals in the proximity of the target animals. In this regard, the fact should be noted that only their infants were situated within arm's range of the target animals. Although no overt social interactions (occurrences of any facial expression or communicative movement) were observed in either the mothers or the infants, the influence of the presence of other group members than the infants did not appear to be a variable affecting this change.

As a possible explanation, one might assume that the mothers were more distracted when the infants were present and thus took longer to clean their teeth than when they were alone. However, the average duration of a given tool-using episode did not increase significantly when the infants were present. More importantly, the mothers' attempts to clean their teeth (as assessed by the number of times they removed the hair per episode) did not increase either when the infants were present. Actually, the average number of hair removal per episode when the infants were present was even smaller than that when the infants were absent. Rather, the change of the pattern of tool-using should be interpreted as a behavioral modification produced by the presence itself of the infants who were watching the mothers.

Concerning human mother-infant interactions, a series of experiments have revealed the fact that strikingly similar parental modifications in their actions, called motionese, can help infants to detect the meaningful structure of the actions [5], [6]. On the basis of observations of 51 and 42 mothers, respectively, who were demonstrating novel objects to their own infants whose ages ranged from 6 months to 13 months, it was found that the mothers tended to modify their infant-directed actions in various ways. They were likely to repeat the actions, to put longer pauses between actions and to exaggerate actions themselves. Such magnification of the movement or 'looming' has been argued so far to play an important role in educating the attention of human infants by attracting their attention due to the occlusion of other sensory information [7].

Indeed, such reasoning is confirmed by an analysis subsequently undertaken from an infant-like viewpoint by applying a model of saliency-based visual attention to such parental action [8], [9]. That analysis was conducted by scientists specializing in robotics originally for the purpose of investigating how such modifications contribute to the infant's understanding of the action. The results of their analysis showed that the model does not suppose any a priori knowledge about actions or objects used in the actions. Instead, it is able to detect and gaze at salient locations, which stand out from the surroundings because of the primitive visual features, in a scene. The model thus demonstrates which low-level aspects of parental actions are highlighted in their action sequences and could attract the attention of young infants, and also robots. Actually, a more recent experimental study [10] demonstrated infants' preference for motionese compared to adult-directed actions by presenting videos of both types of movement to 6- to

13-month-old infants. In the study, the participants showed evidence of such preferences even when demonstrators' faces were blurred in the videos.

Concerning macaques, unlike humans, there is no evidence for imitation under controlled conditions [3]. If we define imitation as the reproduction of the behavior of a model by an observer [11], most empirical studies have failed to show its occurrence in social groups. This could also be the case for the behavior of the monkeys in the present study. In order to explain the spread of the behavior in the group, therefore, we are forced to assume that animals may learn new behaviors from each other through simpler mechanisms than imitation. A typical instance of such reasoning is that its recipient's attention may be drawn to the environment or an object by the presence or interest of the donor itself, even in the absence of any form of intervention of social learning, for the transmission of cultural information. Under such circumstances, again, the modification of the action by the donor is as crucial as it is in the case of imitation because it profoundly affects the likelihood of the recipient acquiring a new behavior, which must be worked out by the recipient itself. The chance that the recipient's resulting behavior comes to resemble the donor's due to environmental or object constraints appears to be facilitated effectively by such modification of the behavior as we report here, which would eventually result in the population-level phenomenon of that behavior.

Methods

The study group was inhabiting a small city, Lopburi, 154 km north of Bangkok, Thailand. In the center of the city stands the old Buddhist shrine of Prang Sam Yot in an open sandy area of approximately 50×50 m surrounded by three 20-m-wide roads and a railway. The present experiment was undertaken there. The area is included in the home range of the study group, which consisted of roughly 200 animals when the study was conducted in February, 2008. Because tourists often visit the shrine when it is open (between 9 a.m. and 5 p.m.), most of the group members were likely to stay there during this period. However, the study group does not spend night there, but in other woody areas at least 1 km away from the shrine. When the research started in 2004, we confirmed the tool-use in 9 adult female monkeys, who rode on the head of female tourists, pulled out their hair, and used it to "floss" their teeth [4]. Since then, the number of animals in which we have confirmed similar behavior has increased up to 50, all of which are adults.

During the study period, 7 females were rearing their approximately 1-year-old infants (3 males and 4 females). We chose all of these 7 females as target animals for the present study. The observation was undertaken by video-recording (30 frames per second) the tool-use of the adults in the area of the shrine. In order to control the variability of the material for the tool-use, we used hairs from a single type of human hairpiece. To provide the stimuli, on each day of observation, we scattered numerous hairs (approximately 20 cm long) that had been dissociated from the hairpieces around the study area early in the morning and waited for the target animals.

The data collection was undertaken using the focal-animal-sampling method. The collection starts with a focal animal, at least 30 min after than the final food-intake of that animal. When using the stimulus as a tool, the animal at first picks it up from the ground. Whenever such behavior is observed, our video-recording is started. When finishing the tool-use, on the other hand, the animal abandons the stimulus onto the ground, and we operationally defined this sequence of handling activity with the stimulus as the material for the teeth-flossing as an 'episode' of the tool-use.

In order to investigate whether the tool-using activity of a target animal was affected by the presence of other group members who were particularly naïve to the activity, we attempted to record the tool-using 'episodes' of the animal when her infant was present in her proximity and when no other animals were present in her proximity. The criterion was solely whether her infant alone remained present within arm's range as well as within the visual range of the target animal throughout a given episode, both animals being situated in a face-to-face position, or whether no animals remained present within such range throughout another given episode. In all, we were able to record 50 episodes where just her infant remained in the target animal's proximity and 50 episodes where no animals remained in the target animal's proximity. In addition, we recorded another 21 episodes during the study period. In these 21, however, animals other than the infant of the target animal entered into proximity with her during the tool-using activity (18 episodes), or the infant was not visually oriented toward the target animal (3 episodes). Thus, data concerning these cases were not included in further analyses.

The video-recording was performed using two video cameras. One of the two filmed the frontal view of the target animal. The tool-using behavior recorded by the videos was coded online by two highly trained coders independently from one another. They were not told the purpose of the present study. The detailed coding schema was essentially the same as that used in our previous study [12]. Overall interrater agreement was 97%. The other camera monitored the area proximal to of the animal. When the infant of the target animal was present in the proximity, the camera filmed its frontal view so that, by analyzing the videos recorded by this second camera and the camera monitoring the target animal, any occurrence of facial expressions and gestural movements could be recorded in both the infant and of the target animal. The occurrences were assessed again by the two raters.

However, none of them reported any occurrence of such communicative behavior in the target animal or in the infant during any episode.

The research methodology complied with protocols approved by the guidelines (Guide for the Care and Use of Laboratory Primates, Second Edition) of Primate Research Institute, Kyoto University, Japan and the legal requirements of Thailand.

Acknowledgements

We are grateful to the Ministry of Environments of Thailand Government for permitting us to undertake this research. We thank Drs Toshiaki Tanaka and Reiko Koba for their assistance in the data analysis, and Dr Elizabeth Nakajima for reading the earlier version of the manuscript and correcting its English.

Authors' Contributions

Conceived and designed the experiments: NM HK NU KW. Performed the experiments: NM HK. Analyzed the data: NM HK. Wrote the paper: NM.

References

1. Goodall JVL (1963) Feeding behavior of wild chimpanzees: a preliminary report. Symp Zool Soc Lond 10: 39–47.

2. Visalberghi E, Fragaszy DM (1990) Do monkeys ape? In: Parker ST, Gibson KR, editors. Language and intelligence in monkeys and apes. Cambridge: Cambridge University Press. pp. 247–273.

3. Ducoing AM, Thierry B (2005) Tool-use learning in Tonkean macaques (Macaca tonkeana). Anim Cognition 8: 103–113.

4. Watanabe K, Urasopon N, Malaivitimond S (2007) Long-tailed macaques use human hair as dental floss. Am J Primatol 69: 940–944.

5. Brand RJ, Baldwin DA, Ashburn LA (2002) Evidence for 'motionese': modifications in mothers' infant-directed action. Dev Sci 5: 72–83.

6. Brand RJ, Shallcross WL, Sabatos MG, Massie KP (2007) Fine-grained analysis of motionese: eye gaze, object exchanges, and action units in infant- versus adult-directed action. Infancy 11: 203–214.

7. Zukow-Goldring P (1997) A social ecological realist approach to the emergence of the lexicon: educating attention to amodal invariants in gesture and

speech. In: Den-Reed C, Zukow-Goldring P, editors. Evolving explanations of development: ecological approaches to organism-environment systems. Washington, D.C.: American Psychological Association. pp. 199–250.

8. Rohlfing KJ, Fritsch J, Wrede B, Jungamann T (2006) How can multimodal cues from child-directed interaction reduce learning complexity in robots? Adv Robotics 20: 1183–1199.

9. Nagai Y, Rohlfing K (2007) Can motionese tell infants and robots "what to imitate"?. Proceedings of the 4th international symposium in animals and artifacts. San Diego: pp. 299–306.

10. Brand RJ, Shallcross WL (2008) Infants prefer motionese to adult-directed action. Dev Sci 11: 853–861.

11. Chauvin C, Berman CM (2004) Intergenerational transmission of behavior. In: Thierry B, Singh M, Kaumanns W, editors. Macaque societies: a model for the study of social organization. Cambridge: Cambridge University Press. pp. 209–234.

12. Masataka N (1992) Motherese in a signed language. Infant Behav Dev 15: 453–460.

CITATION

Masataka N, Koda H, Urasopon N, and Watanabe K. Free-Ranging Macaque Mothers Exaggerate Tool-Using Behavior when Observed by Offspring. PLoS ONE 4(3): e4768. doi:10.1371/journal.pone.0004768. Copyright: © 2009 Masataka et al. Originally published under the Creative Commons Attribution License, http://creativecommons.org/licenses/by/3.0/

Mouse Cognition-Related Behavior in the Open-Field: Emergence of Places of Attraction

Anna Dvorkin, Yoav Benjamini and Ilan Golani

ABSTRACT

Spatial memory is often studied in the Morris Water Maze, where the animal's spatial orientation has been shown to be mainly shaped by distal visual cues. Cognition-related behavior has also been described along "well-trodden paths"—spatial habits established by animals in the wild and in captivity reflecting a form of spatial memory. In the present study we combine the study of Open Field behavior with the study of behavior on well-trodden paths, revealing a form of locational memory that appears to correlate with spatial memory. The tracked path of the mouse is used to examine the dynamics of visiting behavior to locations. A visit is defined as either progressing through a location or stopping there, where progressing and stopping are computationally defined.

We then estimate the probability of stopping at a location as a function of the number of previous visits to that location, i.e., we measure the effect of visiting history to a location on stopping in it. This can be regarded as an estimate of the familiarity of the mouse with locations. The recently wild-derived inbred strain CZECHII shows the highest effect of visiting history on stopping, C57 inbred mice show a lower effect, and DBA mice show no effect. We employ a rarely used, bottom-to-top computational approach, starting from simple kinematics of movement and gradually building our way up until we end with (emergent) locational memory. The effect of visiting history to a location on stopping in it can be regarded as an estimate of the familiarity of the mouse with locations, implying memory of these locations. We show that the magnitude of this estimate is strain-specific, implying a genetic influence. The dynamics of this process reveal that locations along the mouse's trodden path gradually become places of attraction, where the mouse stops habitually.

Author Summary

Spatially guided behavior and spatial memory are central subjects in behavioral neuroscience. Many tasks have been developed for laboratory investigations of these subjects since no single task can reveal their full richness. Here we turn to the simplest and oldest "task," which involves no task at all: introducing a mouse into a large arena and tracking its free behavior. Traditionally, the test is used for studying emotionality and locomotor behavior, using simple summaries of the mouse's path such as its length and the percent of time spent away from walls. More sophisticated computational analysis of the dynamics of the path enables us to separate visiting behavior at locations into stops and passings. Using this distinction, the mouse's path reveals quantifiable locational memory: the mouse's decision to stop in a location is based on its visiting history there. In some strains of mice, the visited locations gradually become places of attraction where the mouse stops habitually. In other strains, the phenomenon is not evident at all. Such quantifiable characterization of locational memory now enables further exploration of the senses that mediate this type of memory and allows measurement and comparisons across mouse strains and across genetic and pharmacological preparations.

Introduction

In the present study we ask how can a kinematic description of Open-Field behavior lead to an understanding of a mouse's higher cognitive functions. We use the organization of elementary patterns for revealing memory-related phenomena.

Low-level kinematic features such as the animal's instantaneous location and speed are extracted from the tracked paths by using special smoothing algorithms [1]. These have been used to statistically partition the mouse's trajectory into intrinsically defined segments of progression and of staying-in-place (stops, lingering episodes; [2]). In previous work on rats, examination of the spatial distribution of stops revealed the home base-the most preferred place in the environment [3]. The home-base is used by the animal as a reference around which it performs structured roundtrips [4],[5]. The home-base also exerts a constraint on the number of stops per roundtrip: the probability of returning to the home-base is an increasing function of the number of stops already performed by the animal in that roundtrip [6]. The home-base acts as an attractor in 2 ways: first, in the vast majority of cases the animal stops in this place upon visiting it, and second, within a roundtrip, this place exerts a gradually increasing attraction on the rat to return to it. Both forms of attraction imply recognition and memory of home-base location. In the present study, starting with the same trajectory data, we approach the issue of recognition and memory of places in a different way, by examining stopping behavior across all locations in the periphery of the open field.

We accomplish this aim by establishing the history of visits to locations all around the periphery of the arena, where visits are classified as stops or passings. We then determine whether the number of previous visits to a location affects the animal's decision to stop in it. An effect of visiting history on the probability of stopping would imply recognition and therefore locational memory.

We used two inbred strains commonly contrasted for their spatial memory—C57BL/6, which is considered to have good spatial memory, and DBA/2, whose performance is poor (e.g., [7]–[9]; see however [10]), and as a third strain, the recently wild-derived strain CZECHII whose spatial behavior might be less affected by domestication.

This study, which has been part of an ethological analysis of mouse exploratory behavior [11]–[14], provides a high throughput test for locational memory.

Results

Since most activity takes place at the periphery of the circular arena (see Methods), we moved to polar coordinates description of the smoothed path (with (0,0) at the center of the circular arena). As illustrated in Figure 1A, the polar projection of the mouse's path as a function of time was punctuated by stops (black dots) in an apparently sporadic manner. While the mouse's decision to stop at a specific location upon traversing it could be taken randomly we wanted to take a closer look at the possibility that it still depended on the history of visits to that location. For that purpose we first established a record of visits in reference to a location, classifying each visit as a stop or a passing through. We then studied jointly records for all locations, and calculated the probability of stopping during a specific visit to a location as a function of the ordinal number of that visit. A change in this probability across visits would have implied that the decision to stop was influenced by visiting history.

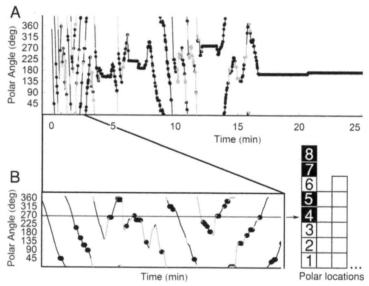

Figure 1. Establishing a Record of Visits in Reference to a Location. (A) CZECHII mouse's polar angles across the first 25 min of a session. Lines represent progression segments and dots represent lingering episodes. The path near the wall is shown in black and the path in the center—in gray. (B) The extraction of a sequence of passings and stops from a time-series of the mouse's polar angles during the first 2.5 min. The horizontal line denotes a specific polar location for which the sequence of visits is extracted, and the numerals printed within squares indicate the ordinal numbers of the visits, white squares for passings, and black—for stops. Only the path near the wall (in black) is used for scoring. The enumerated squares construct, from bottom to top, the column on the right, which depicts the sequence of passings and stops in the selected location.

The procedure of establishing a record of visits in reference to a location is illustrated in Figure 1B: angular position 270° is represented by a straight line parallel to the x-axis. By following the line one can see that upon visiting this location the mouse did not stop in it during the first 3 visits, stopped in it during the 4th and 5th visit, and then again passed through it without stopping during the 6th visit, etc. This sequence of discrete events, consisting of 3 successive passings, 2 stops, 1 passing, and another 2 stops, is presented from bottom to top in the right column of Figure 1B. Similar sequences of passings and stops were obtained for all 120 locations defined by the grid superimposed on the periphery of the arena (see Methods).

The sequences of passings and stops obtained for all locations in 3 representative mouse-sessions are shown in the graphs of Figure 2, left panel. An overview of these graphs reveals that the stops appeared to be distributed evenly throughout the sequences in the DBA mouse, but occurred mostly during later visits in the CZECHII mouse. The increase in stopping frequency across visits was also present in the C57 mouse, but in a milder form. These tendencies appeared to characterize the 3 strains.

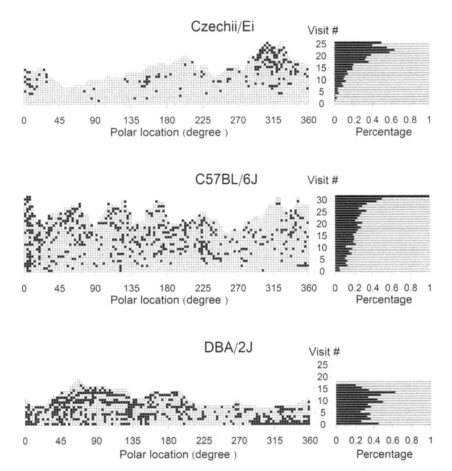

Figure 2. The History of Visits to Peripheral Locations in the Arena. (Left) History of visits to all peripheral locations during a 30-min session of 3 mice belonging to 3 different inbred strains. White squares represent passings, and black squares—stops. (Right) Probability of stopping as a function of the ordinal number of visits. Each horizontal bar represents the proportion of stops performed during the nth visit to a location, by summing up the stops and passings belonging to the corresponding row plotted in the left panel. The black portion of the bar represents the percentage of stops performed during the nth visit to all locations in which such visit occurred (the white portion represents the complementary percentage of passings). As illustrated, the probability of stopping increased as a function of the ordinal number of a visit in the CZECHII and C57 mice, and did not change in the DBA mouse.

We estimated the probability (pn) of stopping during a visit to a location as a function of the ordinal number of that visit in the following way. With Vn being the number of such n-th visits, Sn out of the Vn visits had been classified as stops. The proportion Pn = Sn/Vn was the desired estimator of the probability of interest pn (see Methods). As shown in the right panel of Figure 2, the probability of stopping increased as a function of the ordinal number of a visit in the CZECHII and C57 mice, and did not change in the DBA mouse. In other words, in these CZECHII and C57 mice, the decision to stop in a location was influenced by the number of previous visits paid to that location, whereas in the DBA mouse, visiting history did not affect this decision.

To quantify the rate of change in the probability of stopping, we fitted a linear function of n to the logit-transformed pn in the form (Figure 3):

$$Log\left(\frac{p_n}{1-p_n}\right) = \beta_0 + \beta_1 n,$$

for each mouse. The estimated slopes for all 3 strains are presented in Figure 4. All mice of the CZECHII showed an increase in the probability of stopping as the number of visits increased, so did the trends of all mice of the C57 strain, though the trends were closer to 0. In contrast, DBA showed mixed trends, 21 increasing and 14 decreasing trends. See Figure 4 for the summary of the individual mice trends per each strain and laboratory. Pooling across laboratories using fixed model ANOVA we found that the trend for CZECHII and C57 was significantly positive (p<.0001 and p = .009 respectively) while for the DBA it was not (p = .28) (all results are deposited in the database of the Mouse Phenome Project, [15]).

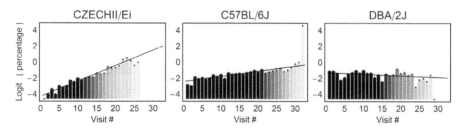

Figure 3. Rate of Change in the Probability of Stopping as a Function of Number of Visits. 3 examples of a linear regression fitted to the normalized probabilities of stopping data. The graphs are similar to the graphs in Figure 2, right panel. Each vertical bar represents the percentage of stops performed during the nth visit to all locations in which such visit occurred. Gray level of bars denotes the weight assigned to the probability value used for the calculation of the linear regression. The data are transformed in order to allow the fitted regression to be linear (see Methods). The black line depicts the regression. The rate of change in the probability of stopping as a function of the ordinal number of a visit was indicated by the slope of the fitted linear function, which reflected a significantly positive trend in CZECHII and C57 mice, and no significant trend in the DBA mouse.

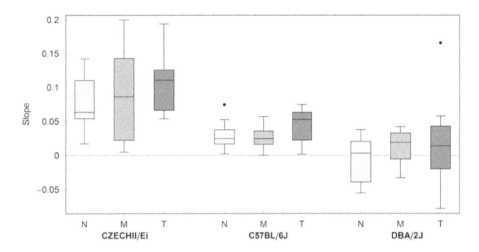

Figure 4. Rate of Change in the Probability of Stopping at a Location. Boxplot summaries of the rate of change in the probability of stopping at a location as a function of the number of previous visits to that location, in 3 strains and across 3 laboratories. Results obtained in NIDA (N), MPRC (M), and TAU (T) are shown, respectively, in light, medium, and dark gray. The trend of the rate of change in the probability of stopping at a location for CZECHII and C57 was significantly positive (p<.0001 and p = .009 respectively) while for the DBA it was not (p = .28).

Putting the result through a more stringent test for replicability, by using the mixed model ANOVA where laboratories were treated as random as well as their interaction with strains [16], we found that the difference in slopes across strains was highly statistically significant (p<.0001). Furthermore, 95% confidence interval for the slope for CZECHII was (.053, .132), for C57 is (−.001, .069) and for DBA was (−.026, .045) giving similar results to those of the fixed effect.

To rule out the possibility that changes in the probability of stopping reflect the level of activity of the animal per session, the Pearson Correlation Test was performed on Distance Traveled near the wall and the slope value obtained from each animal. The correlation was small, r = −.2 and not statistically significant at the .5 level.

The visiting sequences used for the computation of the slopes of regression described the order of visits to the same location; they did not provide the time of the visits' occurrence. The increase in the probability of stopping at locations could, therefore, merely reflect an increase in the frequency of stopping across the session. To examine this possibility we scored the number of stops per sliding time window (3-min time bins with an overlap of 1 min) across the session, fitted a linear regression to the obtained values, and computed the slope of the line. As can be seen in Figure 5, the slopes of all strains in all laboratories were either parallel to the x–axis or negative, implying that the frequency of stopping did not increase across the session. Therefore, changes in the frequency of stopping across time could not explain the change in the probability of stopping at a location with increasing number of visits.

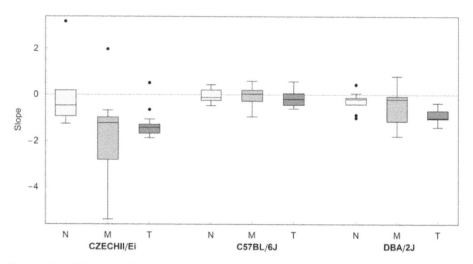

Figure 5. Rate of Change in the Frequency of Stopping across Time. Boxplot summaries of the rate of change in the frequency of stopping across time, in 3 strains tested simultaneously in 3 laboratories. Results, obtained from NIDA (N), MPRC (M), and TAU (T), are shown in light, medium, and dark gray, respectively. The slopes of all strains in all laboratories were either parallel to the x–axis or negative, implying that the frequency of stopping did not increase across the session.

Having ruled out the possibility that the frequency of stopping increases across time, and having shown in the previous section the replicability of the results in 3 laboratories, we concluded that the rate of change in the probability of stopping as a function of visiting history was a reliable measure of mouse locational memory in the open-field.

Dynamics of Stopping in Specific Locations

The changes in the probability of stopping (Figure 4) were computed by pooling the data across all locations at the periphery. Therefore, the results presented so far applied to all locations in a general way, ignoring changes at specific locations. Further investigation of the data collected in TAU revealed 3 types of locations: those in which the probability of stopping increased, those in which it decreased, and those in which it stayed unchanged (see Methods). The locations showing an increase appeared in clusters and so did the locations showing a decrease (see Figure 6). In order to distinguish between the arbitrarily defined single locations, and their clusters, which were revealed by our analysis, we termed the clusters places (it should be noted that the minimal number of locations in a cluster is 3, reflecting our measurement resolution; see Methods).

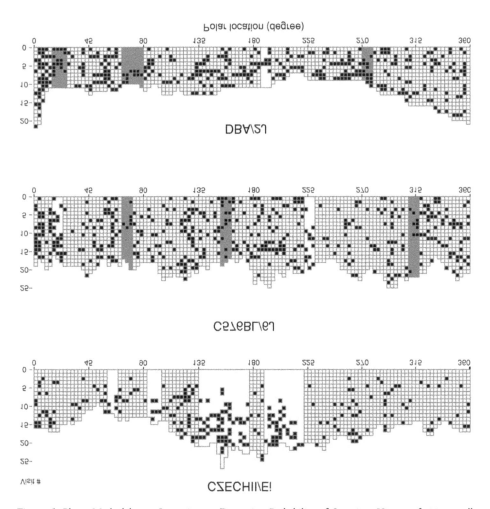

Figure 6. Places Marked by an Increasing or Decreasing Probability of Stopping. History of visits to all peripheral locations across 30-min sessions of 3 mice. White squares: passings, black squares: stops. White stripes mark places in which the probability of stopping increases; gray stripes mark places in which the probability of stopping decreases. The CZECHII mouse was characterized by having only places where the probability of stopping increased or stayed unchanged; in contrast, the C57 and DBA mice were characterized by having all 3 types of places.

As shown in Figure 6 in 3 examples, the CZECHII mouse was characterized by having only places where the probability of stopping increased or stayed unchanged; in contrast, the C57 and DBA mice were characterized by having all

3 types of places; finally, the DBA mouse was characterized by having the highest number of places where the probability of stopping stayed unchanged. Those strain differences prevailed in all the mice tested in TAU.

Discussion

Intrinsic Constraints on Stopping Behavior Imply Locational Memory

In this study we show that in the open field, visiting history to a location influences stopping behavior in that location; the magnitude of this influence is strain specific. In 2 out of 3 examined strains, the higher the ordinal number of a visit to a location, the higher is the probability of stopping in that location. In the third strain, the ordinal number of a visit to a location appears to be irrelevant for the decision whether to pass through the location or stop in it. In the strains that show increased probability of stopping with consecutive visits to a location, this is not due to a general tendency of the mice to stop more frequently with time. On the contrary—the tendency to stop either decreases or stays unchanged across the session in all mice and strains (Figure 5). Because the phenomenon depends on the ordinal number of visits, it implies some type of memory, and because it describes behavior in specific locations, it is spatial. Taken together, it indicates a locational memory.

Locational Memory and Spatial Memory

Future studies would tell us to what extent locational memory utilizes the various sensory modalities. Hippocampus-guided spatial memory is, for example, commonly demonstrated by showing that manipulation of distal visual cues is followed by corresponding adjustments in the animal's spatial orientation [17]. In real life situations, spatial orientation may also be supported by the processing of cues belonging to the other sensory modalities, including proprioception derived from self movement [18], yet the term spatial memory became mainly identified with visual processing. The locational memory highlighted in the present study implies spatial recognition and familiarity, and therefore also reflects spatial memory, but the particular contribution of each of the sensory modalities to the mouse's orientation is not known. Support for visual guidance by distal cues is indicated by the consistency of our results with those obtained for visually guided tests of spatial memory in 2 of the strains (CZECHII mice have not been tested yet for spatial memory). Thus, as with locational memory, good spatial memory is exhibited in C57 in various spatial tasks [9], [19]–[21]. The absence of a locational memory in DBA/2 mice similarly corresponds to the lack of spatial memory reported in most studies performed on this strain [7]–[9], known to suffer from hippocampal dysfunction [22]–[24]. These parallel findings support the hypothesis that the memory described in this paper is also guided visually. The hypothesis that a change in the probability of stopping in locations across visits is mediated by the accumulation of olfactory cues, which are in turn accumulated across visits is untenable, as it would require a mechanism explaining why scent accumulation has no influence on stopping in DBA/2 (Figure 6, lower panel), a strain gifted with a more sensitive olfactory sense than C57 [25],[26], does influence stopping, but in 2 opposite ways, in C57 mice (Figure 6, middle panel), and only increases stopping in the CZECHII mice (Figure 6, upper panel). Estimating locational memory in open field behavior recorded in full darkness would tell us to what extent this construct is supported by information derived from self movement. Finally, dependence on the hippocampus can be investigated by using lesions or temporary inactivation of this structure, and the role played by memory on this phenomenon, although not specific to spatial memory only, can be investigated by using pharmacological disruption that is predictive of memory loss. Whatever the underlying mechanisms, locational memory, which has been shown to be strain-specific can now be compared across strains and preparations.

The Relationship between Locational Memory and the Level of Activity

Since our measure is based on locomotor behavior, there is a concern that this measure is influenced by the animal's level of activity. To rule out this possibility we examined the correlation between distance traveled per mouse-session and the corresponding rate of change in the probability of stopping as a function of the number of visits. The correlation was small and not statistically significant ($r = -.2$, $p<.05$), implying that within the range of values obtained in this study, the level of activity does not influence our measure.

Examination of pharmacological preparations exhibiting hyperactivity (e.g., [27],[28]) could further elucidate the issue of the influence of activity on the measure of locational memory. In 3 previously performed studies on the effect of dopamine-stimulants on locomotor behavior in general, and on stopping in locations in particular, all 3 drugs induced hyperactivity, but had 3 distinct effects on stopping in locations. (+)-amphetamine-induced hyperactivity was associated with a consolidation of stereotypic stopping in a limited number of locations in a relatively fixed order [29]; quinpirole- induced hyperactivity was associated with the performance of stopping in 2 fixed and several varying locations between them [30]; and apomorphine-induced hyperactivity was dissociated from stopping in fixed locations, showing no organization in relation to the environment [31]. Since under the influence of the first 2 drugs the probability of stopping in specific locations increases, locational memory is implied, and our measure would have reflected it. The absence of an increase in the probability of stopping under the influence of the 3rd drug would have resulted in a near-zero rate of change in the probability of stopping implying no locational memory.

Newly Derived versus Classical Inbred Strains

CZECHII mice show a significantly higher rate of change in stopping probability than C57, implying even better spatial abilities. Some researchers consider the behavior observed in classic inbred strains to be dull and "degenerate" [32], whereas wild-mouse behavior is expected to exceed the behavior of these strains [33],[34] in terms of repertoire richness [35], and magnitude of parameters [11],[36]. CZECHII mice are a relatively new wild-derived strain, perhaps less affected by the domestication process; the enhanced spatial performance of this strain could be ascribed to its relative wildness.

A Bottom-Up Approach to Higher Cognition-Related Constructs

The bottom-up approach employed by us aims at revealing higher-level phenomena, as they emerge out of low-level kinematic properties. In the present study, assigning visiting records to locations, and characterizing the sequences constituting these records, reveals locational memory. This phenomenon adds up to a list, reviewed below, of previously described higher-level phenomena also uncovered by the bottom-up approach.

Noting where rats stop, and for how long, highlighted the home-base-the s most preferred place in the arena [3]. Using this place as a reference for measuring kinematic properties of the rat's trajectory revealed several features of the rat's operational world. Partitioning the rats trajectory into roundtrips performed from the home-base highlighted a gradual lengthening of these roundtrips. This lengthening was correlated with an increasing amount of exposure to the arena. It defined, therefore, the animal's increasing familiarity with the environment [5]. A high level of familiarity (= exposure) was also indicated by a reversal of speed differences in relation to the home-base: in a novel environment, the outbound portion of a trip was characterized by lower speeds, and the inbound portion—by higher speeds; in a well-trodden environment the speed difference was shown to be reversed. These speed differences together with the amount of exposure defined "inbound" and "outbound" directions from the rat's point of view [5],[37],[38].

In still another study, the ordinal number of a stop within a roundtrip was found to determine the magnitude of a rat's attraction to the home-base; as the ordinal number of the stop increased, the attraction, expressed as the probability of returning home after stopping, increased as well [6]. Absence of speed differences between inbound and outbound portions were used to infer navigation impairments in hippocampectomized rats [39],[40].

In summary, the dynamics of roundtrip length and of inbound/outbound speed differences were used to define familiarity; the ordinal number of a stop within a trip was used to estimate home-base attraction; and the dynamics of stopping as a function of the ordinal number of visits to locations was used in the present study to estimate spatial memory. The increasing tendency to stop in well-trodden places, in the sense offered by von Uexkull [41], reflects the consolidation of a spatial habit: repeated visits to a location are accompanied by an increasing tendency to stop in that location, culminating in turning it into a relatively stable spatial attractor ([42]; or, in the case of a decreasing tendency to stop, a repeller).

Classifying Locations by their Level of Attraction

A gradual increase or decrease in the probability of stopping along trodden paths reflects respectively an increasing attraction or an increasing repulsion to a location. Whereas the CZECHII mice developed places of attraction and no places of repulsion (Figures 6), the C57 mice (and to an extent also the DBA mice) developed both types. To test the statistical significance of these apparent regularities in single locations, it would be necessary, however, to extend the duration of sessions in order to obtain a much larger number of visits per location.

An Improved Analytical Model of the Kinematic Structure of Rodent Exploratory Behavior

A simple analytical model of rodent exploratory behavior simulated the observations made on real rat open-field behavior [5] by using a sim-rat [37]. The sim-rat increases excursion distance from home-base as a linear function of two system parameters, one governing the rate of motivation loss during movement away from the home-base, and the other the rate of (location-specific) familiarization. The sim-rat's velocity pattern is correlated with the familiarity with places, changing gradually from slow-outbound–fast-inbound, to fast-outbound–slow-inbound. It had been concluded in that analytical study that one shortcoming of the model was that the sim-rat moved continuously, while the movement pattern of a real animal includes stops. It has been further suggested that a comprehensive model of exploratory behavior should include a stochastic component accounting for the stops. The measured changes in the probability of stopping along well-trodden paths specify this stochastic component.

A High Throughput Test of Spatially Guided Behavior in a Less Stressful Environment

The test commonly used for the estimation of spatial memory is the Morris water maze [43]. Other tests include, e.g., the radial arm maze [44], the modified hole board test [45], and the spatial open field [21]. The pros and cons of these and other setups have been discussed elsewhere, and it has been suggested that no single task can reveal the full richness of spatially guided behavior (e.g., [36]). The present study supplements the arsenal of already available tools with a new measure and a new high throughput test of spatially guided behavior conducted in a single session in a large, dry, and empty open field arena.

Methods

The data for this study were collected in a study conducted simultaneously in 3 laboratories: The National Institute on Drug Abuse (NIDA), Baltimore; Maryland Psychiatric Research Center (MPRC), Baltimore; and Tel Aviv University (TAU). These data are stored in a publicly available database (http://www.tau.ac.il/ilan99/see/help), and have already been used in previous studies [11]–[14],[46],[47]. The study included 10 inbred mouse strains and was part of the Mouse Phenome Database project [15]. In this work, we used only the data of the C57BL/6J, DBA/2J and CZECHII/Ei strains.

The experimental and housing protocols were identical for all the above studies, and were described in detail elsewhere [13]. Here we repeat the main points.

Animals

9–14 week old C57BL/6J (C57), DBA/2J (DBA) and CZECHII/Ei (CZECHII) males shipped from Jackson Laboratories. The sample sizes were 12 per C57BL/6J group in each laboratory, 12 per DBA/2J group in each laboratory, and 6 per CZECHII/Ei group in NIDA, 8 in MPRC, and 12 in TAU.

Housing

Animals were kept in a 12:12 reversed light cycle (Light: 8:00 p.m.–8:00 a.m.), and were housed 2–4 per cage under standard conditions of 22°C room temperature and water and food ad libitum. The animals were housed in their room for at least 2 weeks before the start of the experiment. All animals were maintained in facilities fully accredited by the American Association for the Accreditation of Laboratory Animal Care (AAALAC, MPRC and NIDA) or by NIH Animal Welfare Assurance Number A5010-01 (TAU). The studies were conducted at all 3 locations in accordance with the Guide for Care and Use of Laboratory Animals provided by the NIH.

Experimental Procedure

The arenas were 250 cm diameter (TAU, NIDA) and 210 cm diameter (MPRC) circular areas with a non-porous gray floor and a 50-cm high, primer gray painted, continuous wall. Several landmarks of various shapes and sizes were attached in different locations to the arena wall and to the walls of the room where the arena was located. In particular, one wall of the room was mostly covered in black, and a large dark rectangle of 60×80 cm was painted on each of the 2 adjacent walls. The arena was illuminated with two 40-W neon bulbs on the ceiling, above the center of the arena.

The experiments were conducted during the dark part of the cycle, 1–2 hours after its onset. Each experimental animal was brought from its housing room to the arena in a small opaque box, and placed within it (in a standardized location, near the wall) while still in the box. After 20 seconds the box was lifted, and a 30-min session began. The arena was recorded using a resolution of 25 (TAU) or 30 (MPRC, NIDA) samples per second and approximately 1 cm. The animal's movement was tracked using Noldus EthoVision automated tracking system [48].

Data Analysis

The raw data obtained from the tracking system were smoothed using a specialized algorithm implemented in the stand-alone program "SEE Path Smoother" [13],[49]. This procedure produces reliable estimates of momentary speeds during motion (momentary speeds during arrests were defined as zero).

As was previously shown, rodent locomotor behavior consists of two distinct modes of motion—progression segments and lingering episodes [2],[6]. During progression segments, the animals traverse relatively large distances attaining relatively high speeds. During lingering episodes the animals stop and perform scanning movements, while staying in a circumscribed neighborhood. Segmentation of the smoothed path into progression segments and lingering episodes was done using the EM algorithm [50] with a two-gaussians mixture model. Stand-alone user-friendly software for smoothing (SEE Path Smoother) and for segmentation (SEE Path Segmentor) can be downloaded at http://www.tau.ac.il/ilan99/see/help.

Defining Sequences of Visits

Because the vast majority of locomotor behavior is performed along the wall [12],[14] we focused on the path traced by the mouse near it. To quantize the path into sequences of visits to locations, we first schematically superimposed a circular grid consisting of 7×10 cm rectangles on the periphery of the arena. We then partitioned the path traversed by the animal near the wall into a sequence of visits to the locations defined by the grid rectangles. A visit to a location started when the mouse entered the location and ended when it left the location. Because the locations had been defined in an arbitrary way, small insignificant trespassing of the path into adjacent rectangles would have been considered as visits. Therefore, 2 successive visits to the same rectangle were considered as such only if the mouse reached a "long enough" distance from that rectangle between the 2 visits. A "long enough" distance was defined as the distance necessary in order to enter a location that is not adjacent to the original location (see Figure 7). In order to fully surround each location with adjacent locations, an inner-layer of 7×7 cm rectangles was added (the length of the side of the inner-layer rectangles was set to be the same as the width of the outer-layer rectangles).

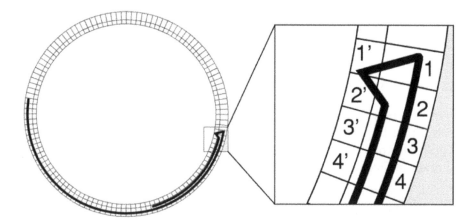

Figure 7. Schematic Illustration of the Partitioning of the Path into Visits. The circle represents the open-field arena with the superimposed grid, consisting of 2 layers of locations. The outer layer was used to define visits to locations, and the inner layer was used to define additional adjacent locations used to avoid false fractionation of visits to outer-layer locations. Insert: outer-layer locations are indexed by numerals, and their corresponding adjacent inner-layer locations are indexed by corresponding numerals with an apostrophe. A black line represents the path traced by the mouse. In this example, 2 visits were scored in locations 3 and 4, 1 visit in location 2, and no visit in location 1.

In the next stage of the analysis, all visits were categorized as either stops (visits containing a lingering episode) or passings (visits that did not contain a lingering episode).

Statistical Methods

Computation of the Slopes of Linear Regression Fitted to the Normalized Probabilities of Stopping

We estimated the probability (p_n) of stopping during a visit to a location as a function of the ordinal number of that visit in the following way. Visits at locations are classified according to whether they constitute an n-th visit to a location or not. Let V_n be the number of such n-th visits (obviously each n-th visit is at a different location); such Vn can be calculated for any n. Out of the V_n n-th visits S_n have been classified as stops. The proportion $P_n = S_n/V_n$ is the desired estimator of the probability of interest p_n.

As often happens when studying the dependence of probabilities on explanatory variables, the dependency of p_n on n seems to follow a logistic model. Namely, we fit a linear function of n to the logit-transformed p_n in the form

$$Log\left(\frac{p_n}{1-p_n}\right) = \beta_0 + \beta_1 n$$

In this model β_0 is the intercept and $\beta 1$ is the slope: β_1 captures the change in the logarithm of the odds $p_n/(1-p_n)$ for a stop, from the n-th visit to the (n+1)-th visit (or equivalently the odds changes gradually from one visit to the next).

Since the variance of Pn as an estimator of pn increases as V_n decreases, being based on a smaller sample, and the latter obviously happens because during late visits the mouse visited increasingly fewer locations, the fitting of the logistic model is based on weighted regression with weights proportional to V_n.

Comparing Endpoint Results Between Strains and Across Laboratories

In order to assess the discrimination between strains and the replicability across laboratories of slopes of logistic regression (see the results section below), we used the linear mixed effects ANOVA model [51],[52]. In this model, the strain was considered as a fixed factor while the effect of laboratory was considered as random. This means that we thought of the laboratory effect as being drawn from the population of all possible laboratories effects. The interaction between strain and laboratory was considered random as well. Thus, a significant strain difference yielded by the Mixed effects Model ANOVA can be regarded as replicable across laboratories. This approach is more conservative than the widely used linear fixed effects model ANOVA: if a difference between two strains was found to be significant under the mixed model, it will be significant under the fixed effects model as well, but the opposite is not necessarily true [16].

One DBA/2 mouse who did not travel along the whole circumference of the arena even once during the session, was excluded from the analysis.

Classifying Locations by the Change in the Probability of Stopping

To investigate changes in the probability of stopping at specific locations, one would have to record many more visits per location than can be collected during a 30-min session. Therefore, we increased sample sizes by pooling the visits paid to 3 adjacent locations at a time, moving along the periphery of the arena with a step of 1 location. To examine the change in the probability of stopping within each group of selected locations we divided each sequence of visits to these locations into 2 halves. When the number of visits was uneven, the visit in the middle of the sequence was excluded from the analysis. We then compared the number of stops and the number of passings in the first half, to their sums in the second half and classified the locations according to the change in the probability of stopping in them. In order to determine whether there was a significant change in the number of stops we used the Fisher Exact Test.

Acknowledgements

We thank Noldus Information Technology for the use of their EthoVision system. We thank Dr. Dina Lipkind for her useful comments on the manuscript.

Authors' Contributions

Conceived and designed the experiments: AD IG. Performed the experiments: AD. Analyzed the data: AD YB. Contributed reagents/materials/analysis tools: YB. Wrote the paper: AD YB IG.

References

1. Drai D, Kafkafi N, Benjamini Y, Elmer G, Golani I (2001) Rats and mice share common ethologically relevant parameters of exploratory behavior. Behav Brain Res 125: 133–140.

2. Drai D, Benjamini Y, Golani I (2000) Statistical discrimination of natural modes of motion in rat exploratory behavior. J Neurosci Meth 96: 119–131.

3. Eilam D, Golani I (1989) Home base behavior of rats (Rattus norvegicus) exploring a novel environment. Behav Brain Res 34: 199–211.

4. Tchernichovski O, Benjamini Y, Golani I (1996) Constraints and the emergence of "free" exploratory behavior in rat ontogeny. J Motor Behavior 133: 519–539.

5. Tchernichovski O, Benjamini Y, Golani I (1998) The dynamics of long-term exploration in the rat. Part I. A phase-plane analysis of the relationship between location and velocity. Biological Cybernetics 78.

6. Golani I, Benjamini Y, Eilam D (1993) Stopping behavior: constraints on exploration in rats (Rattus norvegicus). Behav Brain Res 53: 21–33.

7. Bolivar VJ, Flaherty L (2003) Assessing autism-like behaviors in inbred strains of mice.

8. Rossi-Arnaud C, Ammassari-Teule M (1998) What do comparative studies of inbred mice add to current investigations on the neural basis of spatial behaviors? Exp Brain Res 123: 36–44.

9. Upchurch M, Wehner JM (1988) Differences between inbred strains of mice in Morris water maze performance. Behav Genet 18: 55–68.

10. Wahlsten D, Cooper SF, Crabbe JC (2005) Different rankings of inbred mouse strains on the Morris maze and a refined 4-arm water escape task. Behav Brain Res 165: 36–51.

11. Fonio E, Benjamini Y, Sakov A, Golani I (2006) Wild mouse open field behavior is embedded within the multidimensional data space spanned by laboratory inbred strains. Genes Brain Behav 5: 380–388.

12. Horev G, Benjamini Y, Sakov A, Golani I (2006) Estimating wall guidance and attraction in mouse free locomotor behavior. Genes Brain Behav. In press.

13. Kafkafi N, Lipkind D, Benjamini Y, Mayo CL, Elmer GI, et al. (2003) SEE locomotor behavior test discriminates C57BL/6J and DBA/2J mouse inbred strains across laboratories and protocol conditions. Behav Neurosci 117: 464–477.

14. Lipkind D, Sakov A, Kafkafi N, Elmer GI (2004) New replicable anxiety-related measures of wall versus center behavior of mice in the Open Field. J Appl Physiol. Epub ahead of print.

15. Paigen K, Eppig JT (2000) A mouse phenome project. Mammalian Genome 11: 715–717.

16. Kafkafi N, Benjamini Y, Sakov A, Elmer GI, Golani I (2005) Genotype-environment interactions in mouse behavior: a way out of the problem. Proc Natl Acad Sci USA 102: 4619–4624.

17. O'Keefe J, Nadel L (1978) The hippocampus as a cognitive map. Oxford University Press.

18. Werner S, Krieg-Bruckner B, Mallot HA, Schweizer K, Freksa C (1997) Spatial cognition: the role of landmark, route, and survey knowledge in human and robot navigation. Informatik 41–50.

19. Ammassari-Teule M, Caprioli A (1985) Spatial learning and memory, maze running strategies and cholinergic mechanisms in two inbred strains of mice. Behav Brain Res 17: 9–16.

20. Rossi-Arnaud C, Fagioli S, Ammassari-Teule M (1991) Spatial learning in two inbred strains of mice: genotype-dependent effect of amygdaloid and hippocampal lesions. Behav Brain Res 45: 9–16.

21. Roullet P, Lassalle JM (1990) Genetic variation, hippocampal mossy fibres distribution, novelty reactions and spatial representation in mice. Behav Brain Res 41: 61–70.

22. Paylor R, Baskal L, Wehner JM (1993) Behavioral dissociations between C57BL/6 and DBA/2 mice on learning and memory tasks: a hippocampal-dysfunction hypothesis. Psychobiology 21: 11–26.

23. Thinus-Blanc C, Save E, Rossi-Arnaud C, Tozzi A, Ammassari-Teule M (1996) The differences shown by C57BL/6 and DBA/2 inbred mice in detecting spatial novelty are subserved by a different hippocampal and parietal cortex interplay. Behav Brain Res 80: 33–40.

24. Wehner JM, Sleight S, Upchurch M (1990) Hippocampal protein kinase C activity is reduced in poor spatial learners. Brain Res 523: 181–187.

25. Mihalick SM, Langlois JC, Krienke JD (2000) Strain and sex differences on olfactory discrimination learning in C57BL/6J and DBA/2J inbred mice (Mus musculus). J Compar Psychol 114: 365–370.

26. Wysocki CJ, Whitney G, Tucker D (1977) Specific anosmia in the laboratory mouse. Behavior Genetics 7: 171–188.

27. Geyer MA, Russo PV, Masten VL (1986) Multivariate assessment of locomotor behavior: pharmacological and behavioral analyses. Pharmacology Biochem Behav 25: 277–288.

28. Ralph RJ, Paulus MP, Fumagalli F, Caron MG, Geyer MA (2001) Prepulse inhibition deficits and perseverative motor patterns in dopamine transporter

knock-out mice: differential effects of D1 and D2 receptor agonists. J Neurosci 21: 305–313.

29. Eilam D, Golani I (1994) Amphetamine-induced stereotypy in rats: its morphogenesis in locale space from normal exploration. In: Hendrie CA, editor. Ethology and Psychopharmacology. New York: John Wiley. pp. 241–265.

30. Szechtman H, Sulis W, Eilam D (1998) Quinpirole induces compulsive checking behavior in rats: a potential animal model of obsessive-compulsive disorder (OCD). Behav Neurosci 112: 1475–1485.

31. Szechtman H, Ornstein K, Hofstein R, Teitelbaum P, Golani I (1980) Apomorphine induces behavioral regression: a sequence that is the opposite of neurological recovery. In: Usdin E, Sourkes TL, Youdim MBH, editors. Enzymes and Neurotransmitters in Mental Disease. New York: John Wiley. pp. 511–517.

32. Garland TJ (2003) Selection experiments: an underutilized tool in biomechanics and organismal biology. In: Bels VL, Gasc J-P, Casions A, editors. Vertebrate Biomechanics and Evolucion. Oxford (United Kingdom): BIOS Scientific Publishers. pp. 23–56.

33. Austad SN (2002) A mouse's tale. Natural History 111: 64–70.

34. Silver ML (1995) Mouse Genetics—Concepts and Applications. New York: Oxford University Press.

35. Guenet JL, Bonhomme F (2003) Wild mice: an ever-increasing contribution to a popular mammalian model. Trends Genet 19: 24–31.

36. Wahlsten D, Metten P, Crabbe JC (2003) A rating scale for wildness and ease of handling laboratory mice: results for 21 inbred strains tested in two laboratories. Genes Brain Behav 2: 71–79.

37. Tchernichovski O, Benjamini Y (1998) The dynamics of long-term exploration in the rat. Part II. An analytical model of the kinematic structure of rat exploratory behavior. Biol Cybern 78: 433–440.

38. Tchernichovski O, Golani I (1995) A phase plane representation of rat exploratory behavior. J Neurosci Meth 62: 21–27.

39. Wallace DG, Hines DJ, Whishaw IQ (2002) Quantification of a single exploratory trip reveals hippocampal formation mediated dead reckoning. J Neurosci Meth 113: 131–145.

40. Whishaw IQ, Hines DJ, Wallace DG (2001) Dead reckoning (path integration) requires the hippocampal formation: evidence from spontaneous exploration and spatial learning tasks in light (allothetic) and dark (idiothetic) tests. Behav Brain Res 127: 49–69.

41. Von Uexkull J, Kriszat G (1934) Streifzuge durch die Wumwelten von Tieren und Menschen. Berlin: Springer.

42. Thelen E, Schoner G, Scheier C, Smith LB (2001) The dynamics of embodiment: a field theory of infant perseverative reaching. Behav Brain Sci 24: 1–34.

43. Morris R (1984) Developments of a water-maze procedure for studying spatial learning in the rat. J Neurosci Meth 11: 47–60.

44. Olton DS (1987) The radial arm maze as a tool in behavioral pharmacology. Physiol Behav 40: 793–797.

45. Ohl F, Roedel A, Binder E, Holsboer F (2003) Impact of high and low anxiety on cognitive performance in a modified hole board test in C57BL/6 and DBA/2 mice. Eur J Neurosci 17: 128–136.

46. Kafkafi N, Mayo C, Drai D, Golani I, Elmer G (2001) Natural segmentation of the locomotor behavior of drug-induced rats in a photobeam cage. J Neurosci Methods 109: 111–121.

47. Kafkafi N, Pagis M, Lipkind D, Mayo CL, Benjamini Y, et al. (2003) Darting behavior: a quantitative movement pattern designed for discrimination and replicability in mouse locomotor behavior. Behav Brain Res 142: 193–205.

48. Spink AJ, Tegelenbosch RA, Buma MO, Noldus LP (2001) The EthoVision video tracking system—a tool for behavioral phenotyping of transgenic mice. Physiol Behav 73: 731–744.

49. Hen I, Sakov A, Kafkafi N, Golani I, Benjamini Y (2004) The dynamics of spatial behavior: how can robust smoothing techniques help? J Neurosci Meth 133: 161–172.

50. Everitt BS (1981) Finite. London: Chapman & Hall.

51. McCulloch C, Searle S (2001) Generalized, linear and mixed models. New York: Wiley.

52. Neter J, Kutner M, Nachtsheim C, Wasserman W (1996) Applied linear statistical models. Chicago: Irwin.

CITATION

Dvorkin A, Benjamini Y, and Golani I. Mouse Cognition-Related Behavior in the Open-Field: Emergence of Places of Attraction. PLoS Comput Biol 4(2): e1000027. doi:10.1371/journal.pcbi.1000027.

Altered Behavior and Digestive Outcomes in Adult Male Rats Primed with Minimal Colon Pain as Neonates

Jing Wang, Chunping Gu and Elie D. Al-Chaer

ABSTRACT

Background

Neonatal colon irritation (CI; pain or inflammation) given for 2 weeks prior to postnatal day 22 (PND22), causes long-lasting functional disorders in rats that can be seen 6 months after the initial insult. This study looked at the effect of varying the frequency and duration of neonatal CI on the rate of growth, digestive outcomes, exploratory activity, and colon and skin sensitivity in adult rats.

Methods

Male Sprague-Dawley rats were given CI using repeated colorectal distension (CRD) at different time intervals and for varying durations starting at PND 8, 10 or 14. Control rats were handled by the investigator without any intra-colonic insertion. Further experiments were done on adult rats. Digestive outcomes (food and water consumption, fecal and urinary outputs) were measured using metabolic cages. Exploratory behavior was measured using digital video tracking in an open field. Cutaneous sensitivity was assessed by measuring the responses to mechanical and heat stimuli applied to the shaved abdomen or hind paws. Visceral sensitivity was measured by recording electromyographic responses, under light isoflurane anesthesia, from the external oblique muscles in response to CRD.

Results

No significant weight differences were observed between CI and control rats. Exploratory behavior was reduced in rats with neonatal CI compared to control. Digestive outputs and somatic and visceral sensitivity changed between different treatment groups with earlier and more frequent insults yielding a higher deviation from normal.

Conclusion

The diversity of behavioral and digestive symptoms in these rats parallels the diversity of symptoms in patients with functional gastrointestinal disorders and is consistent with global plastic changes affecting more than one system in the organism.

Background

Level of pain sensitivity, efficacy of analgesics, and susceptibility to developing chronic pain conditions are all subject to individual variability. The sources of such variability can be organismic, environmental or related to personal life history. Organismic variables, such as gender, age, hormonal status, genetic variability, and interactions among these factors have been carefully studied [1-3]. Environmental factors that are extrinsic to the individual such as the social environment, stressful conditions, or light cycle have also been studied and shown to contribute to variability in pain-related traits [4,5]. However, the individual's life history with noxious stimuli has often been overlooked as a potential source of variability in an adult pain experience; yet it is one that can affect sensorimotor processing, pain sensitivity and other behavioral outcomes.

Pain experience early in life can, theoretically, shape the developing nervous system at a time when heightened plasticity characterizes early postnatal development. Chronic abdominal pain is frequently seen during infancy and is often associated with functional constipation. Functional constipation occurs in otherwise well infants at the time of weaning from breast milk to infant formula. Stools become firm with the transition to formula. If a child experiences pain in the anal sphincter while passing a large hard bowel movement, the child becomes conditioned to avoid defecation [6]. The propagating colonic contractions push against an obstructed anal sphincter with pressures of 80 mm Hg and more, well above the threshold for colorectal pain [7]. Colonic painful distension in neonatal rats mimics the naturally occurring pressure build-up in the descending colon and rectum of human infants and provides a reproducible and controllable nociceptive visceral procedure. Most studies have looked at the adult sequelae of child stress and abuse and have focused on emotional and psychological problems, defects in interpersonal relationships, sexual maladjustment and social function [8-10]. Few studies have suggested an association between a history of physical abuse and functional gastrointestinal disorders [11], particularly irritable bowel syndrome (IBS) [12-14]. On the other hand, a number of studies have suggested that sometimes well-intended but painful medical procedures on neonates, without or with inadequate anesthesia, can have negative long-term implications and can engender unwanted consequences [15,16]. In a cohort matched case-control study, using siblings as controls, noxious stimulation caused by gastric suction at birth was associated with an increased prevalence of functional intestinal disorders in later life, possibly linked to the development of long-term visceral hypersensitivity and cognitive hypervigilance [17].

Experimentally, repetitive exposure of neonatal rats, over a period of two weeks, to colon pain using colorectal distension or colon inflammation with mustard oil caused long-term visceral hypersensitivity measurable six months after the initial injury [18]. This hypersensitivity was associated with central and peripheral neural sensitization [19]. Gastric suctioning during the neonatal period also resulted in global chronic somatic and visceral hyperalgesia in adult rats [20]. Similarly, exposure to painful foot-shock in the pre-weanling period had a long-term effect on the sensitivity of rats to painful events [21]. In adult rats exposed to a brief period of inflammation just after birth, the skin receptive field supplied by individual dorsal horn neurons decreased by more than 30% [22], implying permanent alterations in the spinal pain processing for these areas. Short-lasting local inflammation (produced by injection of 0.25% carageenan), produced a long-term hypoalgesia at baseline, which occurred equally in the previously injured and uninjured paws [23]. However, after re-inflammation, a long-term hyperalgesia occurred in the neonatally-injured paw, indicating a significant segmental involvement in the spinal processing of pain [24]. Despite some discrepancies in the results of these studies, most of them indicate that brief or repetitive pain exposures during early periods of development can have a long-term effect on the behavior of the adult.

This study focuses on the effect of variability in the individual's life history on adult digestive function and experience of pain. It examines the effect of neonatal painful colon distension, applied with varying onset time, duration and frequency, on the rate of growth, digestive outcomes (food and water consumption, fecal and urinary output), spontaneous exploratory behavior and visceral and somatic sensitivity of adult rats. Preliminary results were previously reported in abstract form [25].

Methods

Animals

Experiments were done using male Sprague-Dawley rats obtained as pre-weanling neonates (younger than 6 days) from Harlan Sprague-Dawley Inc. (Indianapolis, Indiana). They were housed in plastic cages containing corn chip bedding (Sani-Chips, PJ Murphy Forest Products, Montville, NJ) and maintained on a 12:12 h light:dark cycle (lights on at 07:00 h). The irritation procedure and the experimental testing were conducted during the light component of the cycle. The neonates were housed 10 in a cage with 1 adult female until they were 25 days old. The adult female had access to food and water ad libitum. After separation, the male rats were housed 4 in a cage with access to food and water ad libitum. At the weight of 250 g, only two rats from the same testing group (i.e. control or mechanically irritated) were together in any cage. All studies were performed in accordance with the proposals of the Committee for Research and Ethical Issues of the International Association for the Study of Pain [26] and were approved by the Institutional Animal Care and Use Committee at the University of Arkansas for Medical Sciences in accordance with the guidelines provided by the National Institutes of Health, USA.

Neonatal Colon Irritation

Neonatal rats (8 – 21 days) were exposed to mechanical irritation of the colon of variable duration and at different ages according to the following protocol:

Male Sprague-Dawley rats (8 days old) were divided into 2 groups for purposes of different treatments. Group 1 received colorectal distension (CRD) between the ages of 8 and 21 days at different time intervals and for varying durations and frequencies. The distension was applied using angioplasty balloons (Advanced Polymers Inc., length: 20.0 mm; diameter: 3.0 mm) inserted rectally into the descending colon. The balloon was distended with 0.3 ml of water, exerting a pressure of 60 mmHg (as measured with a sphygmomanometer), for 1 minute and then deflated and withdrawn. The distension was repeated 2 times (separated by 30 minutes) within an hour. This group is referred to as the group with neonatal colon irritation (CI).

Group 2 was handled in a way similar to group 1 except that no colonic insertion was made. In this group, rats between the ages of 8 and 21 days were separated from their mothers for periods of time equal to the corresponding maternal separation in CI rats, but they were only gently held and touched on the perineal area. Group 2 served as control.

The neonatal time period during which the 2 groups were irritated or handled, as well as the duration and frequency of the irritation protocol was varied consistently among the two groups in order to establish a timeline for the onset of long-lasting colon hyperalgesia in these rats. For onset variation, CI was started either on PND8, PND10, or PND14 (PND21 was tested earlier [18]). Duration indicates how long (number of days) the irritation was repeated: 1, 3, or 7 days (we reported earlier the effect of a daily 14 day irritation protocol [18]). Frequency indicates how often the irritation was given, and it varied from daily to every other day. During this period, rats from each group were housed in cages with their mothers. No treatment, procedure or further intervention was done by the investigator for 4 weeks after PND21.

Food and Water Intake and Changes in Fecal and Urine Outputs

Fecal and urine output collections as well as water and food consumptions were evaluated using metabolic cages with wire mesh bottoms (Nalge Company, NY). Adult rats were individually housed (1 rat/cage) for 5 days (2 days for acclimation to the metabolic cage and 3 days for fecal collection). The final fecal output data is the average of fecal collection in the last 3 days. Fecal pellets were collected into a plastic bottle located at the bottom of the cage. The cage is designed to prevent water content in fecal pellets from evaporation and also to prevent urine from getting mixed with feces. The fecal pellets were collected daily, immediately weighed, then baked on a hot plate for two and half hours (95°C) until completely dried. The dried fecal pellets were weighed again. The water percentage contained in the fecal pellets was calculated as follows: [(Weight of fecal pellet before baking) − (weight after baking)]/(Weight of fecal pellet before baking).

The method was further standardized using the following two steps: 1) adjusting the output weight to the rat body weight by dividing the output weight by the rat body weight and obtaining thereby an output weight per gram body weight; and 2) multiplying the (output weight)/(gram body weight) for each rat by a factor of 280 which represents the median weight of all the rats used. Food and water intake were quantified by measuring the amount of food (weight of pellets in g) and water (volume in ml) consumed over a defined period of time and adjusting it to body weight. Besides fecal output, daily urine discharge (ml), daily water (ml) and food (g) consumption, and daily rat body weight were also measured to monitor the daily digestive status of the rats. The data collected from individual CI animals were compared to control; those that fell within control range were described as normal; those that fell outside the control range were described as decreased or increased.

Behavioral Testing

Testing Spontaneous Exploratory Behavior

To measure spontaneous exploratory behavior, we used 4 separate activity enclosures or arenas (San Diego Instruments, CA). The arenas (each 50×50 cm^2) were made of opaque plastic. A PC-linked digital video camera was mounted above the arenas. Each arena was virtually divided into 16 zones using SMART software (Panlab, Spain). Four main parameters were measured: 1) total distance traveled across the arena (cm), 2) average maximum speed at which the animal traveled, 3) number of entries the animal made into a different zone, and 4) total resting time. The enclosures were thoroughly cleaned both before and after each testing period. Animals were always tested at the same time during the day (10 a.m.) in a separate room where no other people or animals were present, with a low noise level and controlled temperature (70–72°F). The activity was recorded over a period of 2 hours for each rat. However, rats tended to become idle after 45 minutes in the arena; therefore, the data was analyzed at 5 minute time point intervals for the first 45 minutes and then compared between CI and control rats.

Testing Somatic Sensitivity

All somatic sensitivity experiments were conducted by an investigator blinded to the type of rat (control or CI). Somatic sensitivity was assessed by measuring the paw or abdomen withdrawal latency to radiant heat as a measure of secondary heat hyperalgesia according to the protocol of Hargreaves et al. [27], or the response to von Frey hair stimulation as a measure of mechanical hyperalgesia. To quantitatively assess the nociceptive threshold to radiant heat of the hind paw, animals were placed in clear plastic cages on an elevated glass plate. The rats were allowed to acclimate for 30 min before testing. A mobile radiant heat source located under the glass was focused onto the hind paw of the rats. The source focused a high intensity light beam through the glass plate onto the plantar surface of the hind paw until the rat lifted its paw. The paw-withdrawal latency (PWL) was recorded by a digital timer. Both hind paws were tested independently (five trials per side; 5-min intervals between trials). The withdrawal latencies for the left and right paws were averaged independently, and the mean value was used to indicate the sensitivity to noxious heat stimulation. The apparatus was adjusted at the beginning of each individual rat study so that the baseline PWL was approximately 10 seconds (s). This setting (i.e., the light beam intensity) was kept unchanged for the remainder of the study. The cut-off of 30 s was used to prevent potential tissue damage.

For abdominal withdrawal latency (AWL), the abdomens of tested rats were shaved and a protocol similar to the one described for PWL was adopted. The light beam was shone on a point on a previously marked area of the lower abdomen, right above the virtual intersection of two imaginary lines extending from the sternum caudally and the lowest rib ventrally. Similar to visceral sensitivity, individual data from CI rats were compared to the mean data from the control group plus or minus two standard deviations (SDs). Individual rats were considered hypersensitive to heat, if PWL or AWL was shorter than the control mean PWL or AWL, respectively, minus two SDs.

The threshold for mechanical hyperalgesia was measured by using a series of calibrated von Frey hairs (Semmes-Weinstein, Stoelting, IL). The plantar surface of the hind paw was touched with different von Frey hairs with a bending force of 0.217–12.5 g. Ten trials were done on each paw. If the rat responded to the stimulation by withdrawing the paw 5 times out of the 10 trials then it was taken as a threshold. If the rat responded to the stimulation by withdrawing the paw more than five times, the next weaker hair was used until the threshold was found. To avoid excessive stimulation, the testing was started in the following sessions with the weakest hair that had elicited withdrawal responses in the previous session. Mechanical hyperalgesia was determined by comparing the number of withdrawal responses out of 10 trials in CI rats versus the number of withdrawal responses in the control rats. A similar protocol was adopted in testing mechanical hyperalgesia on the abdomen.

Testing Visceral Sensitivity

Behavioral responses to colorectal distension (CRD) were assessed in all groups around the age of 3 months by measuring the visceromotor reflex (VMR). The VMR is a reflex measured using an electromyographic (EMG) recording obtained from the external oblique muscle. VMR was recorded in adult rats sedated by light isoflurane anesthesia (2%). While sedated, the rats exhibited no voluntary movements but showed reflexive responses to nociceptive stimuli. The EMG electrode (Teflon-coated stainless steel wire) was inserted through a small skin incision into the external oblique muscle superior to the inguinal ligament and was connected to an amplifier. The signal was displayed on an oscilloscope and fed into a computer using CED 1401 plus and was recorded using Spike 2 software. The raw EMG signal is biphasic and was therefore rectified using "Rectify" script in Spike 2. The rat colons were distended for 20 s every 4 min, and the rectified signal was integrated over the 20 s stimulation period to give a mean response frequency. The mean frequency of the preceding 20 s was subtracted from the distension value to give the EMG intensity as data points. Since these are multiunit recordings, the magnitude of the EMG varied between animals. The data from each animal were normalized to a baseline response derived from the mean of 3 distensions for each intensity of CRD. All subsequent data collected from the same animal were compared to their baseline values. Measuring the threshold intensity of CRD consisted of recording the stimulus intensity that evoked a VMR recorded electronically. CRD was applied in increments of 10 mmHg starting at 10 mmHg (the smallest distinguishable mark on the sphygmomanometer gauge). Comparisons of the responses of CI rats were made to those of control rats in 2 ways: 1) the average response of a group of rats with similar neonatal treatment was compared to the average response of control rats, and 2) the individual normalized data of each rat was compared to the average response of control rats to determine whether the CI rat was hypersensitive, hyposensitive or normo-sensitive. Individual rats whose responses fell within two SDs of the control mean were considered normal. Rats whose responses were smaller than the control mean minus two SDs [< (Mean − 2SD)] were considered hyposensitive, and those whose responses were greater than the control mean + two SDs [> (Mean + 2SD)] were considered hypersensitive.

Rats tested with somatic stimuli were also tested with visceral stimuli and vice versa. The two types of stimuli were separated by 4 days to avoid interactive effects between stimuli.

Colon Stimulation in Adult Rats

Colon stimulation consisted of graded CRD produced by inflating a balloon inside the descending colon and rectum. The balloon was 4 cm in length and made of the finger of a latex glove. It was attached to polyethylene tubing and inserted through the anus into the rectum and descending colon. The open end of balloon was secured to the tubing with thread and wrapped with tape (1 cm wide). The balloon was inserted so the thread was approximately 1 cm proximal to the anal sphincter. The balloon was held in place by taping the tubing to the tail. The tubing was attached via a T connector to a sphygmomanometer pump and a pressure gauge. Prior to use, the balloon was blown up and left overnight so that the latex stretched and the balloon became compliant. CRD was produced by rapidly inflating the balloon to the desired pressure (20, 40, 60 or 80 mmHg) for a duration of 10 s. Stimuli were consecutive (spaced by 20 s) and applied in an ascending graded manner (e.g. 3 × 20, 3 × 40 etc.).

Statistical Analysis

The data were analyzed for statistical significance using Sigma stat software. A Friedman's test was used to assess if responses changed across pressures within each group. The differences in the median values of the EMG response among the 2 treatment groups (CI and Control) at each pressure of CRD were compared using the Kruskal-Wallis (K-W) One Way Analysis of Variance on Ranks. If the K-W test was significant ($p < 0.05$) we did pairwise comparisons using a Wilcoxon Rank Sum test with a Bonferroni correction at 0.05/2 to correct for multiple comparisons. Stated significant results refer to a $p < 0.05/2$.

A One Way Analysis of Variance was made to compare the differences in the median values of the thresholds to elicit a distinctive abdominal contraction measured in the 2 groups. This was followed by pairwise comparisons using a Bonferroni t-test with a corrected p-value of 0.05/2.

For exploratory behavior and somatic sensitivity testing, the responses were compared using a One Way Analysis of Variance between CI and normal rats. Significance was determined whenever $p < 0.05$.

Results

Minimal neonatal CRD in rats leads to a state of mechanical visceral and referred somatic heat hypersensitivity in adults, manifested respectively by increased contractility of abdominal muscles in response to CRD and shorter paw and abdomen withdrawal latencies in response to radiant heat. This state of hypersensitivity exists in the absence of colon inflammation. Tissue examination of sections from the colons of 24 rats (CI, n = 18; Control, n = 6) in various subgroups showed no significant structural damage or loss of crypts. Mucin depletion or increase in intraepithelial lymphocytes was not seen in any of the tissues examined. Slides from the various subgroups were rated as 1+. No significant difference in weight was seen among the adult CI subgroups and between adult CI rats and control rats of similar age. On the other hand, alterations in fecal output could be seen in a number of rats.

Varying the duration and onset time of the neonatal stimulus affected the outcome in adults: a stimulus with a later onset (after PND14) and slower frequency was less likely to evoke all the symptoms seen with an early-onset stimulus (before PND14). However, combining an early-onset with high-frequency stimulation increased the risk for undesirable side effects in the developing rat; these included loss of the animal (death) before it reached adult age, apparent damage to the anal sphincter and occasional enlargement and swelling of the colon seen during histological examination. On the other hand, reducing the frequency of the neonatal CI to one time only, even though given on PND8, did not always produce consistent hypersensitivity in adult rats and the margin of error was larger. In general, rats that received neonatal CI only once showed functional outputs (digestive and behavioral) within the normal range. Optimal combinations of onset-time and duration consisted of a stimulus given twice a day on PND8, PND10 and PND12 or PND10, PND12 and PND14 (see Tables 1, 2, 3).

Table 1. Summary of digestive, behavioral and histological data obtained from adult rats with neonatal CI (onset of CI at PND8).

TABLE I	PND8 (n = 8)	PND8, 9, 10 (n = 12)	PND8, 10, 12 (n = 44)	PND8 – PN14 (n = 12)
Survival	8	10	44 (100%)	8
Food Consumption	1 I; 7 N	4 I; 1 D; 5 N	21 I; 23 N	2 I; 1 D; 5 N
Water Consumption	8 N	3 I; 7 N	8 I; 36 N	3 I; 1 D; 4 N
Dry Feces	2 I; 8 N	5 I; 5 N	21 I; 23 N	4 I; 1 D; 3 N
Water in Feces	8 N	2 I; 3 D; 5 N	6 I; 11 D; 27 N	2 I; 2 M; 4 N
Urine Output	8 N	1 I; 9 N	7 I; 37 N	3 I; 5 N
Somatic Sensitivity	8 N	6 I; 2 D; 2N	22 I; 11 D; 11 N	6 I; 1 D; 1 N
Visceral Sensitivity	2 I; 1 D; 5 N	7 I; 2 D; 1 N	24 I; 12 D; 8 N	5 I; 2 D; 1 N
Colon Histology	N	9 N; 1*	N	5 N ; 3*

Summary of digestive, behavioral and histological data obtained from adult rats with neonatal CI (onset of CI at PND8). Somatic sensitivity refers to the abdominal sensitivity to heat. Visceral sensitivity refers to the responses to CRD (60 mmHg). Colon histology refers to the inflammation grade by comparison to control. N: normal by comparison to the range of control values. I: increased, above the range of normal values. D: decreased, below the range of control values. * indicates animal had colon abnormality (e.g. enlarged colon, relaxed anal sphincter, or other).

Table 2. Summary of digestive, behavioral and histological data obtained from adult rats with neonatal CI (onset of CI at PND10).

TABLE 2	PND10 (n = 8)	PND10, 11, 12 (n = 12)	PND10, 12, 14 (n = 12)	PND10 – PN16 (n = 12)
Survival	8	11	11	9
Food Consumption	8 N	4 I; 2 D; 5 N	5 I; 1 D; 5 N	3 I; 1 D; 5 N
Water Consumption	8 N	3 I; 8 N	2 I; 9 N	5 I; 1 D; 3 N
Dry Feces	1 I; 7 N	5 I; 6 N	4 I; 7 N	4 I; 2 D; 3 N
Water in Feces	8 N	2 I; 3 D; 6 N	3 I; 3 D; 5 N	2 I; 3 D; 4 N
Urine Output	8 N	1 I, 10 N	1 I; 10 N	3 I; 6 N
Somatic Sensitivity	8 N	7 I; 4 N	8 I; 1 D; 2 N	6 I; 2 D; 1 N
Visceral Sensitivity	2 I; 6 N	8 I; 2 D; 1 N	7 I; 2 D; 1 N	7 I; 1 D; 1 N
Colon Histology	N	N	N	5 N; 3*; 1 (lost)

Table 3. Summary of digestive, behavioral and histological data obtained from adult rats with neonatal CI (onset of CI at PND14).

TABLE 3	PND14 (n = 8)	PND14, 15, 16 (n = 12)	PND14, 16, 18 (n = 12)	PND14 – PN20 (n = 12)
Survival	8	12	12	10
Food Consumption	8 N	5 I; 7 N	4 I; 8 N	3I; 2 D; 5 N
Water Consumption	8 N	3 I; 9 N	1 I; 11 N	4 I; 2 D; 4 N
Dry Feces	8 N	3 I; 2 D; 7 N	4 I; 8 N	5 I; 2 D; 3 N
Water in Feces	8 N	3 I; 3 D; 6 N	2 I; 3 M; 7 N	2 I; 3 D; 5 N
Urine Output	8 N	2 I; 10 N	12 N	3 I; 7 N
Somatic Sensitivity	8 N	6 I; 3 D; 3 N	8 I; 4 N	6 I; 4 N
Visceral Sensitivity	1 I; 7 N	5 I; 3 D; 4 N	7 I; 1 D; 4 N	7 I; 3 N
Colon Histology	N	N	N	7 N ; 3*

Changes in Digestive Outcomes

Food and water consumption and fecal and urine outputs were monitored in a total of 149 adult rats: 131 adult rats that received neonatal CRD (see Tables 1, 2, 3) and 18 control rats (12 treated on PND 8, 10 and 12; 6 treated on PND 10, 12 and 14). The results from individual rats with neonatal CI were compared to control rats. For example, the normal range for wet fecal output observed in control rats was between 10 g and 15 g (Fig 1C). Among 44 rats treated on PND8, 10 and 12, twenty-nine rats (65.9%) had normal fecal output (output fell within the normal range), 1 rat (2.3%) had decreased fecal output (between 9 g and 10 g) and 12 rats (12.3%) had increased fecal output (between 15 g and 18 g). The daily percent water in fecal output was within control range (normal) in 27 rats (61.4%), decreased in 11 (25%) and increased in 6 (13.6%) (Fig. 1F). Similar observations were made of other parameters within the same treatment group (Fig. 1) and of other treatment groups (see Tables 1, 2, 3).

Behavioral Studies

Spontaneous Exploratory Activity

Spontaneous exploratory activity was tested in 24 adult rats (control, n = 12; CI, n = 12) that received neonatal treatment on PND8, 10 and 12. Among the 4 parameters measured, the total distance traveled (Fig. 2) was significantly reduced in CI rats compared to control rats. No significant differences were observed in the number of entries from one zone into another (although slightly reduced for CI during the first 25 minutes), their resting time (although it was longer for CI rats than for control rats during the first 10 minutes), or in the maximal velocity of movement between the two groups. Figure 2 illustrates the timeline and the time points at which the distance traveled was measured and shows that CI rats consistently traveled a shorter distance at any time point.

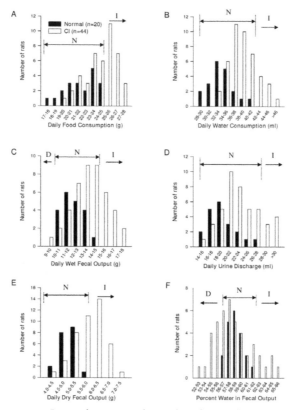

Figure 1. Digestive outcomes. Bar graphs represent the number of rats with a given digestive outcome. The graphs compare 6 measurable outputs of rats treated with neonatal colon irritation (CI; PND8, 10, 12) to control rats (Normal): A. Daily food consumption in g. B. Daily water consumption in ml. C. Daily wet fecal output. D. Daily urine discharge. E. Daily dry fecal output. F. Percent water in fecal output. N: indicates the normal range for each output. D: indicates decreased output (below normal) and I indicates increased output (above normal).

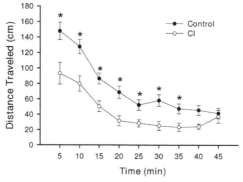

Figure 2. Exploratory activity. Line graphs represent the average distance (+/- SEM) traveled (cm) by control rats (n = 12; solid circle) and rats with neonatal CI (n = 12; open circle), calculated every 5 minutes over a period of 45 min. * means p < 0.05/2.

Somatic Hypersensitivity

Somatic sensitivity was assessed in all groups by measuring the responses to mechanical and heat stimuli applied to the abdomen or the hind paws. No significant differences between CI rats and control rats were observed in the average response to mechanical stimuli, although a slight decrease in the response threshold to abdominal mechanical stimuli was seen in CI rats. By contrast, the average withdrawal latencies to heat stimuli applied to either the abdomen (Fig. 3A) or the hind paws (Fig. 3B) were significantly shorter in CI rats than control rats, indicating that CI rats show signs of heat hypersensitivity referred to the abdomen and the hind paws on both sides. Figure 3 illustrates the reduced average withdrawal latencies to radiant heat in CI rats (PND8, 10 and 12) compared to control rats measured on the abdomen (A) or on the left and right hind paws (B). Similar observations were made in the other CI groups with repeated neonatal CI, although the individual results between groups varied (see Tables 1, 2, 3, somatic sensitivity). In addition, this increased sensitivity to heat stimulation, although significant, was not consistent among all rats of the same treatment group. A number of individual rats showed no change or reduced sensitivity (increased latency) to heat stimuli when compared to control – their responses fell within the range of control responses – but on average, rats that received repeated neonatal CI (3 times or more) were more sensitive than control rats.

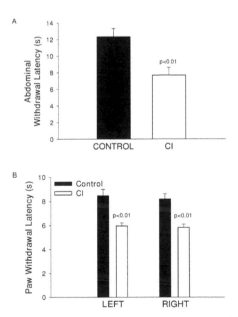

Figure 3. Somatic sensitivity. Bar graphs show the average (+/- SEM) withdrawal latency (s) to radiant heat shone on a shaved area of the abdomen (A) or on the hind paws (B) in control rats (n = 12) and rats with neonatal CI (PND8, 10 and 12; n = 22).

Visceral Hypersensitivity

One hundred and thirty (130) adult rats (age 3 months) treated with neonatal CI were tested for colon hypersensitivity by measuring their responses to CRD using EMG measurement of the abdominal muscle contractions. A majority of adult rats treated with neonatal CI showed visceral hypersensitivity (increased responses to CRD compared to control), with some showing reduced sensitivity and others normal sensitivity (see Tables 1, 2, 3, Visceral Sensitivity). Except for adult rats treated with a single neonatal CI, the average EMG responses to CRD were higher in CI rats than in control (Fig. 4). This visceral hypersensitivity was most significant in rats that received repeated neonatal CI starting at PND8 or PND10. Repetition of neonatal CI for at least 3 times, (e.g. on PND8, PND10 and PND12) yielded a more dramatic increase in visceral hypersensitivity than a single episode (e.g. PND8). However repetition of neonatal CI for more than 3 days (e.g. PND8-PND14) often caused a number of undesirable effects including a higher incidence of pup death. Less significant hypersensitivity was noted with later onset and fewer number of neonatal CI (e.g. onset on PND14 or 1 time CI on PND8).

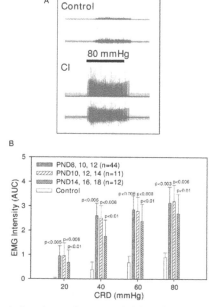

Figure 4. Visceral sensitivity. A. Sample waveforms recorded from the external oblique muscles in a control rat (upper panel) and a rat with neonatal CI (lower panel) in response to CRD (80 mmHg). In each panel, the lower trace represents the raw waveform and the upper trace represents the rectified one. B. Bar graphs represent the average (+/- SEM) EMG intensity (AUC: area under the curve) recorded in response graded CRD (20, 40, 60 and 80 mmHg) in control rats (n = 12) and in adult rats with neonatal CI given at different time points. All data are compared to control. Significance (or lack thereof) is indicated by the value of p shown for each graph. p < 0.05/2 is significant.

When individual responses to CRD of adult rats treated with neonatal CI were compared to the range of individual responses to CRD among control rats, variations in individual visceral sensitivity (increased, decreased or normal) were seen in all groups (see Tables 1, 2, 3, visceral sensitivity). Optimal combinations of onset date and duration were found to be: a) stimulus started on PND8 and repeated on PND10 and PND12; or, b) stimulus started on PND10 and repeated on PND12 and PND14. These combinations of onset and duration minimize the risk of loss of the developing rats while maximizing the significance of relevant symptoms they express as adults compared to controls. For example, adult rats that received neonatal CI at PND8, PND10 and PND12, had a 100% survival rate, increased visceral and somatic sensitivity in addition to a full spectrum of fecal output ranging from reduced to normal to increased (see Tables 1, 2, 3).

Visceral sensitivity correlated with somatic sensitivity but not with digestive outcomes in rats treated with neonatal CI. For example in rats that received neonatal CI on PND8, 10 and 12, responses to CRD (60 mmHg) correlated with the AWL with a correlation coefficient of 0.64 (Fig. 5). Similar correlations were observed between the AWL and responses to CRD (40 and 80 mmHg) for this and other groups. But, the correlation was weak between the PWL and the responses to CRD with maximum $r2 = 0.34$ observed between CRD (60 mmHg) and the PWL in the PND8, 10, 12 group.

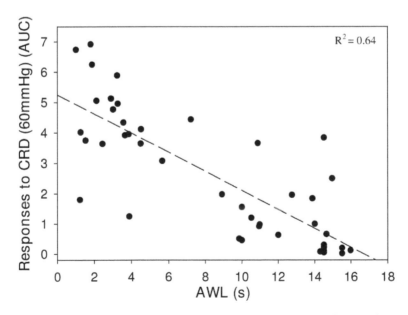

Figure 5. Correlation of visceral and somatic sensitivity. Linear regression curve illustrates the correlation between somatic (AWL) and visceral (CRD = 60 mmHg) sensitivity in CI rats. R2 = 0.64.

Discussion

The main observation made in this study is that limited neonatal irritation to the colon in rats can lead to significantly increased responses to CRD in adults and significantly shorten withdrawal latencies in response to heat stimulation of the abdomen or the hind paws. These observations are consistent with chronic visceral hypersensitivity and increased somatic sensitivity and mimic to a large extent symptoms of chronic abdominal pain with pain referred to somatic structures in humans. In addition, the neonatal insult can reduce exploratory behavior in adult rats and alter their fecal outcome, despite the absence of colon inflammation, mimicking thereby symptoms of discomfort and altered stools in some patients with IBS. These sensory and digestive functional outcomes are, on average, significantly higher in rats with neonatal CI than in control rats. A closer look at the results from individual rats with neonatal CI reveals that, compared to control, the outcomes vary from one animal to the next despite the uniformity of the neonatal insult. This variability is seen in behavioral responses to visceral and somatic stimuli, and also in digestive outputs. For example, when neonatal rats were exposed to the same amount of colon pain, some of them became hypersensitive adults, others became hyposensitive and some did not show any changes in sensory sensitivity. Similarly, fecal output and other digestive parameters increased in some adult rats exposed to neonatal pain, but decreased or stayed within normal ranges in some others. The study offers evidence of individual variability in the pain response and for the first time the possibility of modeling a syndrome of variable symptoms in rats.

Another cardinal observation of this study is that the effects of the neonatal injury were global. They involved somatic and visceral sensory processes (which affect pain-related behaviors), arousal/affective processes (which affect spontaneous exploratory behavior), and digestive processes (which affect fecal and urinary outputs). These observations indicate that a number of plastic changes have taken place in more than one segment of the central nervous system and more likely in more than one system in these rats.

Impact of Varying the Neonatal Injury on Dgestive Outcomes

The best way to describe the effects of repeated neonatal CI on digestive outputs would be as an expansion of the range of outputs beyond those of control animals. The digestive outputs of rats with neonatal CI were analyzed by comparing individual data from the CI rats to the range of normal data obtained from control rats. This approach yielded three different groups of CI rats – rats with normal, increased or decreased outputs – and preserved the statistically significant differences in the average digestive outputs between the groups. If the 3 groups were lumped together, the averaged data would mask individual variability and would show a small, albeit sometimes significant, (one-directional) increase in digestive outputs among all CI rats, instead of the bidirectional shift observed in reality.

This bi-directional shift in outcomes may be caused by a number of factors including: 1) structural or genetic differences between the individual rats that render some rats more or less susceptible to a particular injury or insult – such differences have been known to differentially affect behavioral outcomes [28]; 2) different social ranks among the rats of the same colony (in the same cage) that would render the dominant rats less stressed than the less dominant ones and may affect their behavior and digestive outcomes [29]; and, 3) alternating increase and decrease in outputs of the same rat that are difficult to detect when taking a momentary snapshot. These alternations may be caused by a number of factors, among them altered neural control of GI function at all levels of the nervous system (enteric, peripheral and central) or altered stress response caused by plastic changes in the hypothalamic-pituitary-adrenal axis. Pain in neonates is known to be always stressful [30], and neonatal stress can have a severe impact on adult behavior including increased fecal output, anti-social behavior, neuropsychiatric disorders and other problems in the heart and the gastrointestinal systems [31].

On the other hand, the increases in dry fecal output and in urine output correlated with increased food and water consumption and no change in weight. These correlations seem intuitively simple but are also indicative of a global effect of the neonatal injury. Despite the localization of the neonatal injury to the colon, its functional effects are trans-systemic and are seen throughout the GI tract and urinary system.

Impact on Pain-Related Behaviors

Varying the onset-time, duration or frequency of the neonatal insult differentially affected visceral and somatic sensitivity. Early onset and increased duration and frequency caused the rats to become more hypersensitive and their responses to colorectal distension to become more vigorous. Furthermore, their sensitivity to visceral stimuli correlated well with their sensitivity to somatic stimuli, indicating the involvement of a central mechanism in this sensitivity [18]. These observations were aggravated further the earlier the insult was begun and tended to become less pronounced with later onsets of injury; thus confirming earlier observations that injuries begun after PND21 are unlikely to cause any long-term deficits in neural, behavioral or other functional outcomes [18]. These long-term behavioral changes are related in part to global plastic changes in the sensory systems at more than one level. In fact, plastic neural changes involving adult rats with neonatal injuries have been reported to affect primary afferents innervating the periphery [32], spinal neurons in multiple segments of the cord [18,33]), and thalamic structures and ascending and descending pathways [24,34]. These changes may also affect vagal innervation of the viscera [35], the enteric nervous system and the hypothalamic-pituitary-adrenal "stress" axis [36,37]. The latter may explain the shift in the response to stress in adult rats exposed to neonatal adversity [38-40] and possibly underlies the changes in their digestive outcomes and exploratory behavior.

A distinctive observation of these studies is the hyperalgesia seen in response to heat but not to mechanical stimulation. The lack of mechanical hyperalgesia may be related to the possibility that the mechanical stimulus (probing with von Frey filaments) activates both large and small diameter fibers, which may have been differentially sensitized by the neonatal injury. By contrast heat stimulation activates mainly small unmyelinated fibers. The development of heat hyperalgesia in the hind paws in response to neonatal visceral pain is consistent with observations made in other models of neonatal injury. For example, neonatal gastric suctioning caused visceral hyperalgesia with heat hyperalgesia in the hind paws of adult rats [20]; however, there was no indication of mechanical hyperalgesia or allodynia. In addition, a clear difference has often marked the neural mechanisms of the two types of somatic hypersensitivity. Differentiation between the responses to mechanical and heat stimuli have been widely reported in animal models of pain. For example, NMDA receptors have been shown to mediate heat hyperalgesia but not mechanical allodynia in a rat model of nerve injury; the NMDA antagonist dextrorphan reversed heat hyperalgesia but not tactile allodynia [41,42]. In a recent study using a murine ex vivo somatosensory preparation, the response characteristics of cutaneous sensory neurons staining positively for TRPV1 or TRPV2 were examined. The results suggested that TRPV1 may be essential for heat transduction in a specific subset of mechanically insensitive cutaneous nociceptors and that this subset may constitute a discrete heat input pathway for inflammation-induced thermal pain [43]. Similar differential mechanisms may underlie the sensory divergence in the neonatal colon injury model; however, these remain to be investigated.

The susceptibility of the neonatal organism to painful stimuli may relate to the novelty of the painful stimulus in a neonatal context. The nociceptive neuronal circuits are generally formed during embryonic and postnatal times when painful stimuli are normally absent or limited, but the descending inhibitory pathways, which normally control the flow of nociceptive signals through the spinal cord to higher brain structures, do not mature until later in development [44], leaving the door open for uncontrolled pain signals to wreak havoc in higher brain structures. Therefore, the occurrence of pain during this period is likely to have a more dramatic effect on the plasticity of the nervous system [45,46] and to impact development across several critical time points. It is also likely to be responsible for some of the devastating and permanent behavioral outcomes and disruptions in normal brain activity [47].

On the other hand, the hyposensitivity seen in a subset of neonatally injured rats is in striking contrast with the hypersensitivity seen in a different subset, but consistent with the hyposensitivity reported in other models of neonatal injury [23]. It may be related to a myriad of factors, including 1) desensitization of primary afferents caused by permanent damage during the neonatal period, 2) changes in the sensitization of the pain pathways in the CNS whereby desensitization (similar to long term depression) may have taken effect; or, 3) the triggering of a desensitization response, in what may appear to be an "inoculation" against pain, by neuroimmune factors reported to be involved in the immediate sensitization following injury. It remains to be determined why rats of the same gender, strain and breed respond differently over time to the same neonatal stimulus.

Impact on Spontaneous Behavior

Spontaneous behavior was assessed by measuring exploratory activity in an open field. By comparison to control rats, rats with neonatal CI showed a decrease in the total distance traveled and the number of entries from one virtual zone into another in the open field, and their resting time was longer than that of control rats; however, there was no significant difference in the maximal velocity of movement between the two groups. These observations correlate with reduced exploratory activity. Voluntary exploratory behavior of animals in a new environment may be used as a measure of discomfort that may be associated with ongoing pain [48], distress and anxiety [49], socio-sexual behavior [50], or adaptation to or fear of leaving a familiar place, otherwise known as agoraphobia. The most common psychiatric disorders observed in IBS patients are major depression, panic disorder, social phobia, generalized anxiety, posttraumatic stress and somatization [51]. Walker et al.[52] observed 412 cases of probable IBS (with no other pain problems) and reported that individuals with IBS showed higher rates of major depression (13.4%), panic disorders (5.2%) and agoraphobia (17.8%). Whereas agoraphobia in IBS patients is believed to be caused in most cases by the fear of "not finding a bathroom on time" and can impair a person's lifestyle and career, an organic basis for agoraphobia or an association of agoraphobia with childhood abuse cannot be excluded [53,54]. Studies are underway to determine whether these rats are suffering from depressive or anxiety-like disorders.

Conclusion

Adverse experiences in the neonatal period often contribute to adaptive or mal-adaptive function of the mammalian nervous system, and possibly to the development of chronic intractable disorders that may occur in the absence of identifiable structural problems and which can sometimes be very painful, hence the cognomen 'functional pains' (e.g. fibromyalgia, functional abdominal pain, etc.). These pains can be associated with other functional disorders, can become the consuming focus of a patient's life, and may be onerous to the treating clinician, particularly in the absence of a traceable etiology. This study has shown that an individual's life history with noxious stimuli is a potential source of variability in adult pain experiences and one that can affect sensorimotor processing, pain sensitivity and other behavioral outcomes. In addition, it has shown that a cluster of symptoms as diverse as those observed in irritable bowel syndrome may be modeled in animals. Limited neonatal injury in rats can yield adult outcomes that mimic symptoms of functional GI disorders in man. These symptoms occur in the absence of colon inflammation and include visceral hypersensitivity with referred somatic pain, altered fecal output and reduced exploratory behavior. Whereas the sensitivity can be explained by sensitization of the nervous system, the underlying causes of altered fecal output and reduced exploratory behavior remain to be determined and point to a global phenomenon affecting the organism as a whole rather than just a single system. This is commensurate with the diversity of symptoms in functional GI disorders and the ambiguity of its pathophysiology.

Competing Interests

The authors declare that they have no competing interests.

Authors' Contributions

JW was responsible for the behavioral studies and analysis of related data. CG was responsible for treatment of neonatal rats, digestive outcome studies and analysis of related data. EDA-C was responsible for experimental design, data analysis, statistics, and overall project administration. All authors contributed to the writing of the manuscript.

Acknowledgements

The authors would like to thank Mr. Xin Peng and Ms. Parul Soni for technical assistance, and Ms. Kirsten Garner for administrative assistance.

References

1. Bodnar RJ, Romero MT, Kramer E: Organismic variables and pain inhibition: roles of gender and aging. Brain Res Bull 1988, 21:947–953.

2. Mogil JS, Sternberg WF, Marek P, Sadowski B, Belknap JK, Liebeskind JC: The genetics of pain and pain inhibition. Proc Natl Acad Sci USA 1996, 93:3048–3055.

3. Mogil JS, Chesler EJ, Wilson SG, Juraska JM, Sternberg WF: Sex differences in thermal nociception and morphine antinociception in rodents depend on genotype. Neurosci Biobehav Rev 2000, 24:375–389.

4. Perissin L, Facchin P, Porro CA: Diurnal variations in tonic pain reactions in mice. Life Sci 2000, 67:1477–1488.

5. Perissin L, Facchin P, Porro CA: Tonic pain response in mice: effects of sex, season and time of day. Life Sci 2003, 72(8):897–907.

6. Al-Chaer ED, Hyman PE: Visceral pain in infancy. In Pain in Neonates & Infants. 3rd edition. Edited by: Anand KJS, Stevens BJ, McGrath PJ. Elsevier Churchill-Livingstone; 2007:201–210.

7. Hamid SA, Di Lorenzo C, Reddy SN, Flores AF, Hyman PE: Bisacodyl and high amplitude propagating colonic contractions in children. J Pediatr Gastroenterol Nutr 1998, 27:398–402.

8. Dixon S, Snyder J, Holve R, Bromberger P: Behavioral effects of circumcision with and without anesthesia. J Dev Behav Pediatr 1984, 5(5):246–50.

9. Marshall RD, Schneier FR, Lin SH, Simpson HB, Vermes D, Liebowitz M: Childhood trauma and dissociative symptoms in panic disorder. Am J Psychiatry 2000, 157(3):451–3.

10. Osofsky JD: Neonatal characteristics and mother-infant interaction in two observational situations. Child Dev 1976, 47(4):1138–47.

11. Drossman DA, Camilleri M, Mayer EA, Whitehead WE: AGA technical review on irritable bowel syndrome. Gastroenterology 2002, 123(6):2108–31.

12. Drossman DA, Leserman J, Nachman G, Li ZM, Gluck H, Toomey TC, Mitchell CM: Sexual and physical abuse in women with functional or organic gastrointestinal disorders. Ann Intern Med 1990, 113(11):828–33.

13. Howell S, Poulton R, Talley NJ: The natural history of childhood abdominal pain and its association with adult irritable bowel syndrome: birth-cohort study. Am J Gastroenterol 2005, 100(9):2071–8.

14. Koloski NA, Talley NJ, Boyce PM: A history of abuse in community subjects with irritable bowel syndrome and functional dyspepsia: the role of other psychosocial variables. Digestion 2005, 72(2):86–96.

15. Andrews K, Fitzgerald M: The cutaneous withdrawal reflex in human neonates: sensitization, receptive fields, and effects of contralateral stimulation. Pain 1994, 56:95–101.

16. Fitzgerald M, Millard C, MacIntosh N: Cutaneous hypersensitivity following peripheral tissue damage in newborn infants and its reversal with topical anesthesia. Pain 1989, 39:31–36.

17. Anand KJS, Runeson B, Jacobson B: Gastric suction at birth associated with long-term risk for functional intestinal disorders in later life. J Pediatr 2004, 144:449–54.

18. Al Chaer ED, Kawasaki M, Pasricha PJ: A new model of chronic visceral hypersensitivity in adult rats induced by colon irritation during postnatal development. Gastroenterology 2000, 119:1276–1285.

19. Lin C, Al-Chaer ED: Primary afferent sensitization in an animal model of chronic visceral pain. J Pain 2002, 3(2 Suppl 2):27.

20. Smith C, Nordstrom E, Sengupta JN, Miranda A: Neonatal gastric suctioning results in chronic visceral and somatic hyperalgesia: role of corticotropin releasing factor. Neurogastroenterol Motil 2007, 19(8):692–699.

21. Shimada CKS, Noguchi Y, Umemoto M: The effect of neonatal exposure to chronic footshock on pain-responsiveness and sensitivity to morphine after maturation in the rat. Behav Brain Res 1990, 36:105–111.

22. Rahman W, Fitzgerald M, Aynsley-Green A, Dickenson A: The effects of neonatal exposure to inflammation and/or morphine on neuronal responses and morphine analgesia in adult rats. In Proceedings of the 8th World Congress on Pain. Edited by: Jensen TS, Turner JA, Wiesenfeld-Hallin Z. IASP Press; 1997:783–794.

23. Lidow MS, Song ZM, Ren K: Long-term effects of short-lasting early local inflammatory insult. Neuroreport 2001, 12(2):399–403.

24. Ren K, Anseloni V, Zou SP, Wade EB, Novikova SI, Ennis M, Traub RJ, Gold MS, Dubner R, Lidow MS: Characterization of basal and re-inflammation-associated long-term alteration in pain responsivity following short-lasting neonatal local inflammatory insult. Pain 2004, 110(3):588–96.

25. Ma H, Park Y, Al-Chaer ED: Functional outcomes of neonatal colon pain measured in adult rats. J Pain 2002, 3(2 Supp 1):27.

26. Zimmermann M: Ethical guidelines for investigations of experimental pain in conscious animals. Pain 1983, 16:109–110.

27. Hargreaves K, Dubner R, Brown F, Flores C, Joris J: A new and sensitive method for measuring thermal nociception in cutaneous hyperalgesia. Pain 1988, 32:77–88.

28. Mogil JS, Miermeister F, Seifert F, Strasburg K, Zimmermann K, Reinold H, Austin JS, Bernardini N, Chesler EJ, Hofmann HA, Hordo C, Messlinger K, Nemmani KV, Rankin AL, Ritchie J, Siegling A, Smith SB, Sotocinal S, Vater A, Lehto SG, Klussmann S, Quirion R, Michaelis M, Devor M, Reeh PW: Variable sensitivity to noxious heat is mediated by differential expression of the CGRP gene. Proc Natl Acad Sci USA 2005, 102(36):12938–43.

29. Chesler EJ, Wilson SG, Lariviere WR, Rodriguez-Zas SL, Mogil JS: Identification and ranking of genetic and laboratory environment factors influencing a behavioral trait, thermal nociception, via computational analysis of a large data archive. Neurosci Biobehav Rev 2002, 26(8):907–23.

30. American Society of Pediatrics: Prevention and Management of Pain and Stress in the Neonate. Pediatrics 2000, 105(2):454–461.

31. Barreau F, de Lahitte JD, Ferrier L, Frexinos J, Bueno L, Fioramonti J: Neonatal maternal deprivation promotes Nippostrongylus brasiliensis infection in adult rats. Brain Behav Immun 2005, 20(3):254–260.

32. Lin C, Al-Chaer ED: Sensitization of thoracolumbar primary afferent responses to colorectal distension (CRD) in an animal model of chronic visceral pain. In Program No. 451.10. 2002 Abstract Viewer/Itinerary Planner. Washington, DC: Society for Neuroscience, CD-ROM; 2002.

33. Lin C, Al-Chaer ED: Differential effects of glutamate receptor antagonists on dorsal horn neurons responding to colorectal distension in a neonatal colon irritation rat model. World J Gastroenterol 2005, 11(41):6495–6502.

34. Saab CY, Park YC, Al-Chaer ED: Thalamic modulation of visceral nociceptive processing in adult rats with neonatal colon irritation. Brain Res 2004, 1008(2):186–92.

35. Powley TL, Martinson FA, Phillips RJ, Jones S, Baronowsky EA, Swithers SE: Gastrointestinal projection maps of the vagus nerve are specified permanently in the perinatal period. Brain Res Dev Brain Res 2001, 129(1):57–72.

36. Plotsky PM, Thrivikraman KV, Nemeroff CB, Caldji C, Sharma S, Meaney MJ: Long-Term Consequences of Neonatal Rearing on Central Corticotropin-Releasing Factor Systems in Adult Male Rat Offspring. Neuropsychopharmacology 2005, 30(12):2192–2204.

37. Schwetz I, McRoberts JA, Coutinho SV, Bradesi S, Gale G, Fanselow M, Million M, Ohning G, Tache Y, Plotsky PM, Mayer EA: Corticotropin-releasing factor receptor 1 mediates acute and delayed stress-induced visceral hyperalgesia in maternally separated Long-Evans rats. Am J Physiol Gastrointest Liver Physiol 2005, 289(4):G704–12.

38. Coutinho SV, Plotsky PM, Sablad M, Miller JC, Zhou H, Bayati AI, McRoberts JA, Mayer EA: Neonatal maternal separation alters stress-induced responses to viscerosomatic nociceptive stimuli in rat. Am J Physiol Gastrointest Liver Physiol 2002, 282(2):G307–16.

39. Hinze C, Lin C, Al-Chaer ED: Estrous cycle and stress related variations of open field activity in adult female rats with neonatal colon irritation (CI). In Program No. 155.14. 2002 Abstract Viewer/Itinerary Planner. Washington, DC: Society for Neuroscience, CD-ROM; 2002.

40. Yamazaki A, Ohtsuki Y, Yoshihara T, Honma S, Honma K: Maternal deprivation in neonatal rats of different conditions affects growth rate, circadian clock, and stress responsiveness differentially. Physiol Behav 2005, 86(1–2):136–44.

41. Tal M, Bennet GJ: Extra-territorial pain in rats with a peripheral mononeuropathy: mechano-hyperalgesia and mechano-allodynia in the territory of an uninjured nerve. Pain 1994, 57(3):375–82.

42. Wegert S, Ossipov MH, Nichols ML, Bian D, Vanderah TW, Malan TP Jr, Porreca F: Differential activities of intrathecal MK-801 or morphine to alter responses to thermal and mechanical stimuli in normal or nerve-injured rats. Pain 1997, 71(1):57–64.

43. Lawson JJ, McIlrath SL, Woodbury CJ, Davis BM, Koerber HR: TRPV1 unlike TRPV2 is restricted to a subset of mechanically insensitive cutaneous nociceptors responding to heat. J Pain 2008, 9(4):298–308.

44. Fitzgerald M, Koltzenburg M: The functional development of descending inhibitory pathways in the dorsolateral funiculus of the newborn rat spinal cord. Brain Res 1986, 389(1–2):261–70.

45. Peng YB, Ling QD, Ruda MA, Kenshalo DR: Electrophysiological changes in adult rat dorsal horn neurons after neonatal peripheral inflammation. J Neurophysiol 2003, 90:73–80.

46. Ruda MA, Ling QD, Hohmann AG, Peng YB, Tachibana T: Altered nociceptive neuronal circuits after neonatal peripheral inflammation. Science 2000, 289:628–631.

47. Ren K, Novikova SI, He F, Dubner R, Lidow MS: Neonatal local noxious insult affects gene expression in the spinal dorsal horn of adult rats. Mol Pain 2005, 1(1):27.

48. Palecek J, Paleckova V, Willis WD: The roles of pathways in the spinal cord lateral and dorsal funiculi in signaling nociceptive somatic and visceral stimuli in rats. Pain 2002, 96(3):297–307.

49. Griebel G, Perrault G, Sanger DJ: Limited anxiolytic-like effects of non-benzo-diazepine hypnotics in rodents. J Psychopharmacol 1998, 12:356–65.

50. Gonzalez MI, Albonetti E, Siddiqui A, Farabollini F, Wilson CA: Neonatal organizational effects of the 5-HT2 and 5-HT1A subsystems on adult behavior in the rat. Pharmacol Biochem Behav 1996, 54:195–203.

51. Lydiard RB: Irritable bowel syndrome, Anxiety and depression: what are the links? J Clin Psychiatry 2001, 62(Suppl 8):38–45.

52. Walker EA, Katon WJ, Jemelka RP, Roy-Byrne PP: Comorbidity of gastro-intestinal complaints, depression and anxiety in the Epidemiologic Catchment Area (ECA) Study. Am J Med 1992, 92(Suppl 1A):265–305.

53. Mancini C, Van Ameringen M, MacMillan H: Relationship of childhood sexual and physical abuse to anxiety disorders. J Nerv Ment Dis 1995, 183(5):309–14.

54. Young SJ, Alpers DH, Norlend CC, Woodruff RA: Psychiatric illness and the irritable bowel syndrome: practical implications for the primary care physician. Gastroenterology 1976, 20:162–166.

CITATION

Social Transmission of Avoidance Behavior Under Situational Change in Learned and Unlearned Rats

Akira Masuda and Shuji Aou

ABSTRACT

Background

Rats receive information from other conspecifics by observation or other types of social interaction. Such social interaction may contribute to the effective adaptation to changes of environment such as situational switching. Learning to avoid dangerous places or objects rapidly occurs with even a single conditioning session, and the conditioned memory tends to be sustained over long periods. The avoidance is important for adaptation, but the details of the conditions under which the social transmission of avoidance is formed are unknown. We demonstrate that the previous experience of avoidance learning is important for the formation of behaviors for social transmission of avoidance and that the experienced rats adapt to a change of situation determined by the presence or absence of aversive stimuli. We systematically investigated social influence on avoidance behavior using a passive avoidance test in a light/dark two-compartment apparatus.

Methodology/Principal Findings

Rats were divided into two groups, one receiving foot shocks and another with no aversive experience in a dark compartment. Experienced and inexperienced rats were further divided into subjects and partners. In Experiment 1, each subject experienced (1) interaction with an experienced partner, (2) interaction with an inexperienced partner, or (3) no interaction. In Experiment 2, each subject experienced interaction with a partner that received a shock. The entering latency to a light compartment was measured. The avoidance behavior of experienced rats was inhibited by interaction with inexperienced or experienced partners in a safely-changed situation. The avoidance of experienced rats was reinstated in a dangerously-changed situation by interaction with shocked rats. In contrast, the inexperienced rats were not affected by any social circumstances.

Conclusions/Significance

These results suggest that transmitted information among rats can be updated under a situational change and that the previous experience is crucial for social enhancement and inhibition of avoidance behavior in rats.

Introduction

Various social animals interact with conspecifics and use information from other animals to adapt to their environments. The transmission of information by interaction or observation is called social transmission. Social transmission is shaped by social clues, which consist of visual, olfactory, acoustic, or other types of information from conspecifics. Many studies have shown that social interaction or simple observation of other animals' behavior has significant effects on food preference [1]–[3], acquisition of motor patterns [4]–[6], and avoidance [7]–[9] in many species of vertebrate including primates, birds, fish, and rodents (for a review, see [10]). Rats, one of the most common experimental animals, prefer to ingest the same type of food as that ingested recently by a conspecific [2], [11]. This social transmission of food preference is thought to be formed by an association between food odorants and a volatile component of a rat's breath [12].

One of the most important behaviors affecting survival is the avoidance of dangerous objects or places. Avoidance learning is formed through an operant-conditioning process. In passive avoidance, for example, animals are punished for entering a preferred place by a footshock, and then the animals stop entering the place. This learning also includes some aspects of Pavlovian-conditioning [13]–[14]. In avoidance learning, association between an aversive stimulus and the environmental context (and its components) also can be shaped. Some previous studies reported that rats did not learn avoidances socially [15]–[16]. For example, rats do not learn avoidance just by watching conspecifics receiving a shock [17] or by interaction with poisoned conspecifics [14]. Other paper showed that rats learned to avoid a candle flame by exposure to another rat acquiring the same avoidance responses [18]. These conflicting results probably come from the different experimental conditions.

One possible factor is subjects' experience. The various responses following social interaction could be affected by the responder's experience. For example, social recognition requires semantic memories and knowledge obtained previously by experiences [19]. The perception of another's pain, and empathy for pain, are dependent upon bottom-up factors (i.e., observation of another person's pain expression and contextual pain cues) as well as top-down factors (i.e., features of the observer's own experience of pain and knowledge) (for a review, see [20]).

A recent study has shown that rats, like humans, can apply previous learning to adapt to new situations [21]. Social transmission of food preference also interacts previous learning in rats [22]. Therefore, experience of individual learning should be important for various perceptions and decision making even by rats. In the previous studies concerning social transmission of avoidance, many researchers used naïve rats as subject animals. Considering that not only social cues but also subjects' experience are important for social recognition, we believe one possible explanation why rats did not learn avoidance socially could be that the association between top-down factors (avoiding experience of individuals) and bottom-up factors (social clues from others) was not formed because naïve rats have no experiences of pain or another aversive stimulus.

Adaptive behavior learned in response to a dynamic environment is surely determined by the changing conditions of the environmental situation. There is dynamic interaction between the learning of avoidance behavior and a situation. Avoidance behavior is adaptive in an environment that includes a danger, but this behavior will be discontinued if the danger disappears. Social influence has the potential to improve the adaptation to an environment with a situational change, because the probability of receiving a signal of danger or safety as well as the possibility of sharing the signal change becomes high in social conditions. However, the effect of social influence on adaptation to a change of situation, especially from danger to safety or safety to danger, is not known, while that of adaptation to a novel situation has been investigated in detail. In the present study we focused on subjects' experience of avoidance learning and investigated uncertain dynamics, that is, the social influence on avoidance behavior in response to a situational change. We conducted two sequential experiments. In Experiment 1, we examined the effect of social interaction on avoidance behavior in a safely-changed situation where the shock stimulus was lost. In Experiment 2, we examined the social influence in a dangerously-changed situation where the shock stimulus was renewed.

Materials and Methods

Subjects

The subjects were 77 male Wistar rats aged 8 weeks, acquired from Kyudo Co., Ltd. (Kumamoto, Japan). They were given free access to food and water, and housed two per cage for one week before the start of the experiments. Housing conditions were thermostatically controlled at 22–24°C with a light/dark cycle (lights on: 08:00—20:00). The experiments were performed under the control of the Ethics Committee of Animal Care and Experimentation in accordance with the Guiding Principles for Animal Care Experimentation, Kyushu Institute of Technology, Japan, and with the Japanese Law for Animal Welfare and Care.

Apparatus

The experiments took place in a test chamber consisting of two compartments, a light compartment (D25 cm×W25 cm×H27 cm) and a dark compartment (D30 cm×W30 cm×H30 cm) (Figure 1A). The two compartments were divided by a sliding door. Electric shocks are delivered by a shock generator (SGS-002, Muromachi Kikai Co., Ltd., Tokyo, Japan). In Experiment 2, a removable partition was used to prevent subject animals from moving from one compartment to the next earlier than the partners.

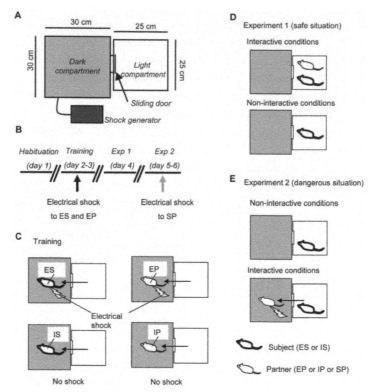

Figure 1. The experimental design. (A) Experimental apparatus. (B) Time schedule of this study. The black arrow shows electric shock to the experienced subjects and partners (ES: experienced subjects; IS: inexperienced subjects; EP: experienced partners; IS: inexperienced partners). The gray arrow shows electric shock to the partners (SP: shocked partners). (C–E) Overview of the experiments. (C) The schematic diagram of the training session. The left row indicates the treatment for the subjects (ES: experienced subjects; IS: inexperienced subjects); the right row indicates the treatment for the partners (EP: experienced partners; IP: inexperienced partners). (D) The schematic diagram of Experiment 1. The upper row indicates interactive conditions, and the lower row indicates non-interactive conditions. (E) The schematic diagram of Experiment 2. The upper row shows non-interactive conditions, and the lower row shows interactive conditions.

Procedure

All treatments or behavioral tests were done during the light cycle (12:00–20:00) in the following sequence (the whole schedule is shown in Figure 1B):

Training

On the first day of this session (day 1), all animals were placed in the light compartment for 1 min individually and habituated to the experimental apparatus. After this interval, the sliding door was raised and the latency to enter the dark compartment was recorded. On the second day (day 2), a single electrical shock

(0.5 mA, 5 s) was induced inescapably on 40 animals in the dark room after each animal entered the dark compartment, and they were used as the experienced subjects and 5 partners. The other 37 animals who received no electrical shocks were used as inexperienced subjects and partners. The experimental apparatus was cleaned with alcohol to remove odors before treating the next subject. On the third day (day 3), the latency of each animal to enter the dark compartment was measured. The schematic diagram of the training is shown in Figure 1C.

Experiment 1

The subjects were divided into three groups: i) together with an experienced partner (EP), ii) with an inexperienced partner (IP), iii) without any partner (No). On the day following the training session (day 4), each subject was placed in the light compartment. If partnered, they were paired with the partner rats for 1 min. After the interval, the sliding door was raised and then the latencies to enter the dark compartment were measured, with a cut-off time of 15 min. This experiment was performed without any electric shocks. The schematic diagram of Experiment 1 is shown in Figure 1D.

Experiment 2

The day after Experiment 1 was performed (day 5), experienced and inexperienced subjects were put in the experimental apparatus individually and habituated to the dark compartment for 20 min. On the second day of this experiment (day 6), 30 min before the test trial, each animal was placed in the light compartment and then the latencies to enter the dark compartment were measured, with a cut-off time of 5 min. We used the experienced subjects that entered within a given cut-off time as the experienced subjects (n = 16) and randomly selected inexperienced subjects (n = 12). In a test trial, each subject was placed in the light compartment with a partner for 1 min. Then, the sliding door was raised to permit the partners only to enter the dark compartment. During this time, a mesh partition attached in the center of the light compartment (in between a subject and a partner) did not permit the subjects to enter the dark compartment. After the partner entered, electrical shocks (0.5 mA, 3–6 s) were induced. Immediately after that, the partner returned to the light compartment and stayed there. After an additional interval (30 s), the partition was removed. The latencies to enter the dark compartment were measured with a cut-off time (15 min). The partner rat stayed in the light compartment and could interact with the subject freely during the measurement. We then compared the latency between the two conditions, with no partner and with a shocked partner. The schematic diagram of Experiment 2 is shown in Figure 1E. All partners were the rats already used in

Experiment 1, which had been given a single foot shock to stabilize partners' pain reaction (habituation to the shock).

Data Analysis

Data were analyzed with the use of SPSS software (version 16.0). Before analysis, the Kolmogorov-Smirnow test was performed for normality. In Experiment 1, we used the Turkey-Kramer multiple comparison test to assess the statistical significance of the difference among the rat groups. In Experiment 2, we used a paired t-test to evaluate the statistical significance of the difference between measurements in the absence and presence of partners. The criterion for statistical significance was $p < 0.05$ (two-tailed).

Results

Social Interaction on Avoidance Behavior in a Safe Situation

For preparation, we trained 40 rats individually (30 subjects and 10 partners) to avoid the dark room by using electrical stimuli (0.5 mA, 5 s), and the other 37 rats (28 subjects and 9 partners) did not receive the training. The trained rats and untrained rats were used as experienced rats and inexperienced rats, respectively. We examined the influences of social interaction on the avoidance behaviors in a safe situation under the following 6 conditions: i) experienced subjects with inexperienced partners (ES-IP), ii) experienced subjects with experienced partners (ES-EP), iii) experienced subjects without any partners (ES-No), iv) inexperienced subjects with inexperienced partners (IS-IP), v) inexperienced subjects with experienced partners (IS-EP), (vi) inexperienced subjects without any partners (IS-No). As a summary, all combinations are presented in Table 1.

Table 1. The conditions for Experiment 1 and Experiment 2.

Condition	Interactive	Subject	Partner
Experiment 1			
(i) ES-IP	YES	Experienced	Inexperienced
(ii) ES-EP	YES	Experienced	Experienced
(iii) ES-No	NO	Experienced	(-)
(iv) IS-IP	YES	Inexperienced	Inexperienced
(v) IS-EP	YES	Inexperienced	Experienced
(vi) IS-No	NO	Inexperienced	(-)
Experiment 2			
ES-No	NO	Experienced	(-)
ES-SP	YES	Experienced	Shocked
IS-No	NO	Inexperienced	(-)
IS-SP	YES	Inexperienced	Shocked

One day after the preparation (day 2), we measured the step-through latency of both subjects of each pair individually. All of the experienced rats refrained from entering the dark compartment within 5 min (mean±s.e.m. = 1102+40 s), while inexperienced rats entered within 1 min (mean±s.e.m. = 15+2 s). The difference among the groups in experienced subjects (p>0.6, for all pairs, Figure 2A) was not significant. A similar result was found in inexperienced subjects (p>0.5, for all pairs, Figure 2B). The next day, we measured the latency with social interaction under the safe condition. We found that the latency of the ES-IP group was significantly shorter than that of the ES-EP (p<0.001, ES-IP vs. ES-EP) and ES-No (p<0.0001, ES-IP vs. ES-No) groups. Interestingly, the avoidance responses of rats in the ES-EP group was also shortened (p<0.01, ES-EP vs. ES-No, Figure 2C). The latencies of all three groups of inexperienced subjects, however, were not different from one another (p>0.8, for all pairs, see Figure 2D). Similar results were found in the staying duration in the dark compartment of both experienced and inexperienced subjects (Figure 2E–F).

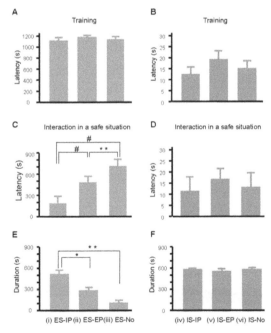

Figure 2. The effect of social interaction on avoidance behaviors in a safe situation. (A) Step-through latency (mean+s.e.m.) of the experienced subjects during the testing performed 24 h after shocking in the dark compartment of the experimental apparatus. (B) The step-through latency of the inexperienced subjects. (C) The latency of experienced subjects after interaction with inexperienced partners (ES-IP), after interaction with experienced partners (ES-EP), and after no interaction (ES-No). (D) The latency of inexperienced subjects under the three conditions (IS-IP, IS-EP, IS-No). (E-F) The duration of staying in the dark compartment. The number of subjects was ES-IP (n = 10$); ES-EP (n = 10); ES-No (n = 10); IS-IP (n = 9), IS-EP (n = 10$), IS-No (n = 9). $: Marked conditions were measured at the same time. The means±s.e.m. are represented as bars. The duration of one IS-IP subject was deleted due to the failure of measurement. (*, p<0.05, **, p<0.01, #, p<0.001)

The Effect of Social Interaction on Avoidance Behavior in a Dangerous Situation

The partners were given a foot shock stimulus during the retention time of the subjects, and we then compared the latency between asocial and social conditions. All the conditions tested are described in Table 1. This behavioral test was conducted using identical animals because the avoidance behavior of the experienced subjects can vary individually. First, we measured the subjects' basal avoidance without social interaction (ES-No, IS-No). The mean latency of ES-No was 123.6±19.4 (s), and that of the IS-No was 8.3±1.6 (s). There was a significant difference between the ES-No and IS-No groups (p<0.001). After a 30-min interval, the subjects were placed in the experimental setting again, where they interacted with shocked partners (ES-SP, IS-SP), and the latencies of the subjects were measured. The latency of the ES was significantly increased by the interaction with shocked partners (ES-No vs. ES-SP, p<0.05, Figure 3A). Not all, but some of them showed clearly prolonged retention. On the other hand, the avoidance behavior was not enhanced in inexperienced subjects at all. Their latency tended to decrease rather than increase (IS-No vs. IS-SP, p = 0.1, Figure 3B). These results indicate that the information from shocked partners had a facilitatory effect on avoidance in the experienced subjects.

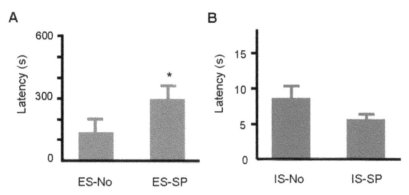

Figure 3. The effect of social interaction on avoidance behaviors of ES and IS in a dangerous situation. (A) Latency of the experienced subjects under an asocial condition (ES-No) and under a social condition (ES-SP). (B) Latency of the inexperienced subjects under a non-interactive condition (IS-No) and under an interactive condition (IS-SP). The numbers of experienced subjects (ES-No and ES-SP) and that of the inexperienced subjects (IS-No and IS-SP) were n = 16 and n = 12, respectively. * p<0.05.

Discussion

In the current study, the behavioral influences of social interaction between two rats in a changing environment were systematically evaluated by focusing on the previous experience of passive avoidance learning. The major results were as follows: (1) learned avoidance behavior was inhibited by social interaction with neighboring partners, especially partners who had not learned avoidance behavior; (2) avoidance behavior of experienced rats was reinstated by shocked partners; (3) there were none of the inexperienced rats whose avoidance behavior was modified by any kind of partner. Taken together, these results indicate that previous learning is a crucial factor for the social enhancement or inhibition of avoidance in rats. Our findings suggest a view in which the prerequisites for the social transmission of avoidance may include previous learning experience of subjects as well as alarming social cues from others.

Social Interaction Induces an Inhibitory Influence on the Avoidance of Experienced Subjects in Safe Conditions

The experienced subjects were inhibited by the partners under the no-shock conditions. These inhibitory influences of social interaction were also found in learned aversion to a flavored food [23] and conditioned fearful response [24]–[25]. The results of this study present that the social interaction has the inhibitory effect also on the passive avoidance in rats. A new finding of the present study is that inhibitory influences depend on a partner's experience. The strength of inhibitory influence was much higher by inexperienced partners than by experienced partners. This suggests that the previous learning of partners has a specific role in the social modulation of avoidance. How do social partners affect avoidance behavior of other individuals? Some studies have shown that individual vigilance was depressed by increasing group size [26]–[27] or by shortening neighbor distance [28] in various animals. The depressed vigilance may prompt an inhibitory influence on avoidance. These effects can explain the inhibitory influence of experienced partners. The group effect cannot explain why the influence of inexperienced partners is higher than that of experienced partners. Inexperienced partners inhibited the avoidance more strongly than did experienced partners, even though the two rats were placed in a very limited space under the ES-EP conditions. Therefore, there are likely other mechanisms at work. One most likely reason why the effect was bigger with the inexperienced partners rather than the experienced partners is that the subjects followed the partners. Rats have been thought to have some high-order cognitive abilities such as imitation through observation of acting others [29]–[30] and causal reasoning [31]. Two other possibilities are: (1) the rats might imitate the behavior of inexperienced partners introduced to the dark compartment without awareness, and (2) the rats might expect extinction of the dangerous stimuli by inference from the partners' behavior. These two possibilities are formed by the independent effect of observation, and it would be necessary to examine this effect to know if these possibilities are feasible.

Social Interaction with Shocked Partners can Induce a Facilitatory Influence on Avoidance

As already mentioned, previous studies suggested that social transmission of avoidance does not occur in naïve rats [13]–[15]. The present results that the avoidance behavior of inexperienced subjects was not facilitated by social interaction under either no-shock conditions or shock conditions are consistent with the results of previous studies. However, we observed that social interaction facilitated avoidance in avoidance-experienced rats. A previous study also showed that conditioned fear was recovered by the presentation of shocked partners in Pavlovian conditioning [32]. Our results provide the possibility of social transmission of avoidance in an operant learning paradigm, that is, not only under Pavlovian conditioning but also operant conditioning. The present systematic experiments empirically showed the unexamined differences between avoidance-related adaptation of experienced rats and inexperienced rats under social environments.

Social Cues for the Social Transmission About Avoidance in Rats

Animals transmit various types of social cues, and those signals tell important information to other companions. The present results clearly demonstrate that social cues from a partner determine the contents of social transmission. Social cues emitted by partner rats can be categorized into two types according to the partners' situations regarding stimulus application. One category of social cue is accompanied by punishment or negative stimulus such as an electrical shock to individual animals. This type of social cue can be an announcement of an aversive situation or danger for others. Actually, for example, fish emit alarm substances when they are attacked by an enemy. Those substances are social cues that trigger avoidance in others [33]–[34]. Another category of social cue is accompanied by reward or positive stimulus such as food to individuals. That can be an announcement of a favorite situation or safety for others. The social transmission of food preference in rats [2] is an example.

What signals are important for the adaptation to a changing environment? In the present study we investigated the social transmission of information with environmental change from danger to safety and vice versa, and our results may help to answer the question. The experimental design allowed partners to have interaction with subjects. In Experiment 2, the partner was able to transmit sensory information including (1) alarming vocalization emitted when the partner was shocked, (2) smell or pheromone in excretion such as urine and feces, and (3) struggling motion. Shock or stress can induce alarming vocalization (ultrasonic vocalization) [35]–[36] and alarming odors [37] in rats. The timing of transmission varies according to the nature of the information. Vocalization and struggling motion tended to be emitted just after the partner was shocked, and then they faded within a few seconds. In contrast, odor information was emitted from the shocked partner after shocking, but it lasted a relatively long time. Therefore, one of those forms of sensory information or a combination of them may have acted as signals to announce danger. For an announcement of safety, the lack of a shock-induced reaction of partners may be an important signal. In social animals, avoiding or facilitating a behavior by many types of social cue effectively controls their adaptation to a changing environment.

How does Individual Experience Affect Social Transmission?

The present results demonstrate that social interaction affects experienced subjects' behavior but not inexperienced subjects' behavior, especially in a dangerously-changed situation (Figure 3). This indicates that there is an experience-dependence of social interaction in avoidance behavior and that previous individual experiences play an important role in social transmission. What is the importance of the learning experience in the processes of social transmission? There seem to be two possibilities, at least. First, individual experiences work to enhance the acquisition of information from other animals during observation. Some studies indicate aversive experiences enhance the sensitivity of animals with respect to the acquisition of information [38]–[40]. Getting information from others is the first step of social transmission. There is no doubt about the importance of the quality of getting information in social transmission. How can this explain the present results? By following this hypothesis, experienced subjects were affected by other partners because of the enhancement of previously gained sensitivity, but inexperienced subjects were not affected because their sensitivity level was not high enough. This interpretation can partially explain the present results, but it is difficult to explain all of the results with only this interpretation for the following reason. In this experiment none of the inexperienced subjects was affected by partners, although inexperienced subjects received similar social cues to those received by experienced subjects. Actually, a previous study has shown that inexperienced rats get information from other conspecifics showing fear responses [41]. This is inconsistent with the first hypothesis, but second hypothesis can explain that result.

A second possible reason for the importance of the learning experience in the processes of social transmission is that when getting social cues, individual experiences are recalled and help the receiver to associate individual experience with information from other conspecifics to plan the next appropriate action. This is another promising hypothesis. If avoidance-learning is recalled under the influence of a partner's cues, avoidance behavior will be quickly reacquired even after avoidance responses are extinct. Although there is no direct evidence that individual memory is recalled via another conspecific in rats, memory can be recalled by various associative stimuli. The neural mechanism where social cues are associated with individual experiences should be elucidated in the future. These two possible functions may support the notion of stages of social transmission.

Our results provide evidence that individual experience is one of the important factors for social enhancement or inhibition of avoidance behavior. It may be that we have little knowledge about the social transmission of avoidance because behavioral experiments focusing on individual experience are not so popular. Additional progress of the behavioral studies considering individual experience may facilitate the understanding of the neural mechanism for social learning through cooperation with researchers conducting neurological studies that have been revealing the neural mechanisms of various types of learning.

Conclusion

In conclusion, we systematically investigated social influence on avoidance behavior under a situational change, focusing on the previous experience of rats. Throughout our experiments, the experienced subjects were influenced by experienced or inexperienced partners depending on changing experimental situations. The results suggest that rats can adapt their behaviors by utilizing both social interaction with a variety of types of partners and individual experiences in dynamically changed situations.

Authors' Contributions

Conceived and designed the experiments: AM SA. Performed the experiments: AM. Analyzed the data: AM. Contributed reagents/materials/analysis tools: AM. Wrote the paper: AM.

References

1. Zentall TR, Levine JM (1972) Observational learning and social facilitation in the rat. Science 178: 1220–1221.

2. Galef BG Jr, Kennett DJ, Stein M (1985) Demonstrator influence on observer diet preference: Effects of simple exposure and the presence of a demonstrator. Anim Learn Behav 13: 25–30.

3. Hikami K, Hasegawa Y, Matsuzawa T (1990) Social transmission of food preferences in Japanes.e.m.onkeys (Macaca fuscata) after mere exposure or aversion training. J Comp Psychol 104: 233–237.

4. Hirata S, Morimura N (2000) Naive chimpanzees' (Pan troglodytes) observation of experienced conspecifics in a tool-using task. J Comp Psychol 114: 291–296.

5. Biro D, Inoue-Nakamura N, Tonooka R, Yamakoshi G, Sousa C, et al. (2003) Cultural innovation and transmission of tool use in wild chimpanzees: evidence from field experiments. Anim Cogn 6: 213–223.

6. Carlier P, Jamon M (2006) Observational learning in C57BL/6j mice. Behav. Brain Res 174: 125–131.

7. Hall D, Suboski MD (1995) Visual and olfactory stimuli in learned release of alarm reactions by zebra danio fish (Brachydanio rerio). Neurobiol Learn Mem 63: 229–240.

8. Brown G (2003) Learning about danger: Chemical alarm cues and local risk assessment in prey fishes. Fish and Fisheries 4: 227–234.

9. Cook M, Mineka S, Wolkenstein B, Laitsch K (1985) Observational conditioning of snake fear in unrelated rhesus monkeys. J Abnorm Psychol 94: 591–610.

10. Galef BG Jr, Laland KN (2005) Social learning in animals: Empirical Studies and Theoretical Models. BioSci 55: 489–500.

11. Galef BG Jr, Whiskin EE (2003) Socially transmitted food preferences can be used to study long-term memory in rats. Learn Behav 3: 160–164.

12. Galef BG Jr, Mason JR, Preti G, Bean NJ (1988) Carbon disulfide: A semiochemical mediating socially-induced diet choice in rats. Physiol Behav 42: 119–124.

13. Mowrer OH (1947) On the dual nature of learning: A re-interpretation of "conditioning" and "problem-solving." Harv Educ Rev 17: 102–148.

14. Baum M (1969) Dissociation of respondent and operant processes in avoidance learning. J Comp Physiol Psycho 67: 83–88.

15. Galef BG Jr, Wigmore SW, Kennett DJ (1983) A failure to find socially mediated taste-aversion learning in Norway rats (R. norvegicus). J Comp Psychol 97: 358–363.

16. Galef BG Jr, McQuoid LM, Whiskin EE (1990) Further evidence that Norway rats do not socially transmit learned aversions to toxic baits. Anim Learn Behav 18: 199–205.

17. White DJ, Galef BJ Jr (1998) Social influence on avoidance of dangerous stimuli by rats. Anim Learn Behav 26: 433–438.

18. Lore R, Blanc A, Suedfeld P (1971) Empathic learning of a passive-avoidance response in domesticated Rattus norvegicus. Anim Behav 19: 112–114.

19. Gallagher HL, Frith CD (2003) Functional imaging of theory of mind. Trends Cogn Sci 7: 77–83.

20. Goubert L, Craig KD, Vervoort T, Morley S, Sullivan MJ, et al. (2005) Facing others in pain: The effects of empathy. Pain 118: 285–288.

21. Murphy RA, Mondragón E, Murphy VA (2008) Rule learning by rats. Science 319: 1849–1851.

22. Galef BG Jr, Whiskin EE (2001) Interaction between social and individual learning in food preferences of Norway rats. Anim Behav 62: 41–46.

23. Galef BG Jr (1986) Social interaction modifies learned aversions, sodium appetite, and both palatability and handling-time induced fietary preference in rats (Rattus norvegicus). J Comp Psychol 100: 432–439.

24. Kiyokawa Y, Kikusui T, Takeuchi Y, Mori Y (2004) Partner's stress status influences social buffering effects in rats. Behav Neurosci 118: 798–804.

25. Kiyokawa Y, Takeuchi Y, Mori Y (2007) Two types of social buffering differentially mitigate conditioned fear responses. Eur J Neurosci 26: 3606–3613.

26. Elgar MA (1989) Predator vigilance and group size in mammals and birds: A critical review of the empirical evidence. Biol Rev 64: 13–33.

27. Lima SL, Dill LM (1990) Behavioral decisions made under the risk of predation: A review and prospectus. Can J Zool 68: 619–640.

28. Elgar MA, Burren PJ, Posen M (1984) Vigilance and perception of flock size in foraging house sparrows (Passer domesticus L.). Behavior 90: 215–223.

29. Heyes CM, Dawson GR (1990) A demonstration of observational learning in rats using a bidirectional control procedure. Q J Exp Psychol 42B: 59–71.

30. Heyes CM, Dawson GR, Nokes T (1992) Imitation in rats: Initial responding and transfer evidence. Q J Exp Psychol Sect B 45: 229–240.

31. Blaisdell AP, Sawa K, Leising KJ, Waldmann MR (2006) Causal reasoning in rats. Science 311: 1020–1022. [DOI: 10.1126/science.1121872].

32. Riess D (1972) Vicarious conditioned acceleration: Successful observational learning of an aversive pavlovian stimulus contingency. J Exp Anal Behav 18: 181–186.

33. von Frisch K (1938) Zur psychologie des fische-schwarmes. Naturwissenschaften 26: 601–606.

34. Magurran AE, Irving PW, Henderson PA (1996) Is there a fish alarm pheromone? A wild study and critique. Proc Roy Soc London Ser B 263: 1551–1556.

35. DeVry J, Benz U, Schreiber R, Traber J (1993) Shock-induced ultrasonic vocalization in young adult rats: a novel model for testing putative anti-anxiety drugs. Eur J Pharmacol 249: 331–339.

36. Blanchard RJ, Blanchard DC, Agullana R, Weiss SM (1991) Twenty-two kHz alarm cries to presentation of a predator, by laboratory rats living in visible burrow systems. Physiol Behav 50: 967–972.

37. Mackay-Sim A, Laing DG (1981) The sources of odors from stressed rats. Physiol Behav 27: 511–513.

38. Fletcher ML, Wilson DA (2002) Experience modifies olfactory acuity: Acetylcholine-dependent learning decreases behavioral generalization between similar odorants. J Neurosci 22: RC201.

39. Wilson DA (2003) Rapid, experience-induced enhancement in odorant discrimination by anterior piriform cortex neurons. J Neurophysiol 90: 65–72.

40. Li W, Howard JD, Parrish TB, Gottfried JA (2008) Aversive learning enhances perceptual and cortical discrimination of indiscriminable odor cues. Science 319: 1842–1845.

41. Knapska E, Nikolaev E, Boguszewski P, Walasek G, Blaszczyk J, et al. (2006) Between-subject transfer of emotional information evokes specific pattern of amygdala activation. Proc Natl Acad Sci USA 103: 3858–3862.

CITATION

Masuda A and Anou S. Social Transmission of Avoidance Behavior under Situational Change in Learned and Unlearned Rats. PLoS ONE 4(8): e6794. doi:10.1371/journal. pone.0006794. Copyright © 2009 Masuda, Aou. Originally published under the Creative Commons Attribution License, http://creativecommons.org/licenses/by/3.0/

Molecular Variation at a Candidate Gene Implicated in the Regulation of Fire Ant Social Behavior

Dietrich Gotzek, D. DeWayne Shoemaker and Kenneth G. Ross

ABSTRACT

The fire ant Solenopsis invicta and its close relatives display an important social polymorphism involving differences in colony queen number. Colonies are headed by either a single reproductive queen (monogyne form) or multiple queens (polygyne form). This variation in social organization is associated with variation at the gene Gp-9, with monogyne colonies harboring only B-like allelic variants and polygyne colonies always containing b-like variants as well. We describe naturally occurring variation at Gp-9 in fire ants based on 185 full-length sequences, 136 of which were obtained from S. invicta collected over much of its native range. While there is little overall differentiation between most of the numerous alleles observed, a surprising amount is

found in the coding regions of the gene, with such substitutions usually caus-
ing amino acid replacements. This elevated coding-region variation may re-
sult from a lack of negative selection acting to constrain amino acid replace-
ments over much of the protein, different mutation rates or biases in coding
and non-coding sequences, negative selection acting with greater strength on
non-coding than coding regions, and/or positive selection acting on the pro-
tein. Formal selection analyses provide evidence that the latter force played
an important role in the basal b-like lineages coincident with the emergence
of polygyny. While our data set reveals considerable paraphyly and polyphy-
ly of S. invicta sequences with respect to those of other fire ant species, the b-
like alleles of the socially polymorphic species are monophyletic. An expand-
ed analysis of colonies containing alleles of this clade confirmed the invariant
link between their presence and expression of polygyny. Finally, our discovery
of several unique alleles bearing various combinations of b-like and B-like
codons allows us to conclude that no single b-like residue is completely predic-
tive of polygyne behavior and, thus, potentially causally involved in its expres-
sion. Rather, all three typical b-like residues appear to be necessary.

Introduction

A main goal of evolutionary genetics is to document naturally occurring variation
and to reconcile observed patterns with population history and demography, fitness consequences, and selection regimes at genes of interest [1], [2]. Study of the
adaptive maintenance of molecular variation historically has followed one of two
approaches, elucidation of the functional components of molecular adaptations at
the biochemical level, or description of the historical footprints of selection acting
on sequence variants [3]. An important objective in modern studies of molecular
adaptation is to bridge the two approaches by means of comprehensive research
integrating functional data with information on patterns of variation that impli-
cate past selection [4] (see also [3], [5]–[7]).

The fire ant Solenopsis invicta displays an important colony-level social polymorphism that is associated with variation at a single gene, general protein-9
(Gp-9) [8]. Colonies with a single reproductive queen (monogyne colonies) always feature the exclusive presence of B-like alleles of Gp-9 in all colony members. In contrast, colonies with multiple reproductive queens (polygyne colonies)
always have an alternate class of alleles, designated b-like alleles, represented along
with B-like alleles among colony members [8]–[11]. This pattern, coupled with
similar genetic compositions of monogyne and polygyne populations at numerous other nuclear genes, has led to the hypothesis that the presence of b-like
alleles in a colony's workers is both necessary and sufficient to elicit polygyne

social behavior [8]–[10], [12], [13]. Because variation in queen number represents a fundamental social polymorphism that is associated with a suit of important reproductive, demographic, and life history differences [14]–[16], variation at Gp-9 is hypothesized to underlie the expression of major alternative adaptive syndromes in S. invicta.

Our understanding of the association of variation at Gp-9 with colony social organization has been advanced by the production of sequence data for S. invicta in its native (South American) and introduced (USA) ranges as well as for numerous other Solenopsis species [10], [11]. These studies revealed several important patterns. First, the monophyletic b-like alleles are restricted to a clade of six South American fire ant species that display the monogyne-polygyne polymorphism (this group of species, which includes S. invicta, is informally termed the socially polymorphic clade). Second, polygyne colonies in all of the socially polymorphic species always contain b-like alleles. Third, the b-like alleles of the socially polymorphic species bear diagnostic amino acid residues at positions 42, 95, and 139 that distinguish them from all other Gp-9 alleles (collectively known as B-like alleles). This finding prompted speculation that the substitutions at one or more of these positions may alter the function of GP-9 protein with respect to its proposed role in modulating social behavior [10], [11]. Finally, the b alleles, which comprise a small clade of b-like alleles that apparently arose recently in S. invicta, feature a radical, charge-changing substitution (Glu151Lys) that may underlie their observed deleterious effects in homozygous condition [17].

GP-9 protein is a member of the insect odorant-binding protein family [10]. Several well studied proteins in this family have been implicated as important molecular components of chemoreception in insects, presumably effecting the transduction of pheromones or food chemostimulants to neuronal signals by transporting these ligands through the chemosensillar lymph to neuronal receptors [18]. Regulation of colony queen number in fire ants involves reciprocal chemical signalling and perception between workers and queens, with workers ultimately making decisions about which queens, and how many, are tolerated as colony reproductives based on queen pheromonal signatures [9], [19]. The role of some odorant-binding proteins in chemoreception, the invariant association of one class of Gp-9 alleles with polygyny, and the restriction of this class of alleles to the socially polymorphic clade of South American fire ants have been viewed as evidence that Gp-9 may directly influence social organization rather than merely being a marker for other genes of major effect on this trait (reviewed in [20]).

Further complexities in our understanding of Gp-9 and fire ant social evolution have arisen as additional sequence data have been generated. For instance, it is now apparent that variation at this gene is not invariably associated with colony social organization in the genus Solenopsis, given that the fire ant S. geminata, a distant North American relative of S. invicta, does not exhibit allelic variation associated with colony social form [21]. Also, discovery of a Gp-9 allele with a b-like amino acid residue at position 95 but B-like residues at positions 42 and 139 in the undescribed S. species "X" indicates that the two classes of alleles in the socially polymorphic species are not as internally homogeneous as previously believed [11]. Unfortunately, the social organization of the source colony for this sequence could not be determined, precluding a test of the importance of residue 95 in regulating queen number. These recent results suggest that progress in our understanding of Gp-9 in fire ants has been hampered by limited knowledge of naturally occurring sequence variation; in even the best studied species, S. invicta, only a handful of individuals from a single site in the native range have been sequenced to date.

To remedy this shortcoming, the present study documents sequence variation at Gp-9 using extensive samples of S. invicta collected over a large portion of its vast native range as well as samples of many of its fire ant relatives. Few previous studies have employed such large collections of intraspecific sequence variants in order to analyze patterns of adaptive molecular variation at single genes [22]. Our specific objectives were to characterize the molecular evolution of Gp-9 in fire ants, to test for effects of selection on the gene, to confirm the association between polygyny and the presence of variants encoding b-like amino acid residues, and to test for phylogeographic patterns in the distribution of the observed variation in S. invicta. We were particularly interested in finding new variants encoding unique combinations of B-like and b-like residues at codons 42, 95, and 139, with the hope that their discovery might shed light on the role of each substitution in mediating the expression of social organization. In combination with the other analyses, information from such variant colonies is expected to aid progress in connecting the genetic and phenotypic variation underlying regulation of fire ant social behavior.

Results

General Results

The complete data set consisted of 185 full-length Gp-9 sequences (149 newly generated), of which 136 were from the focal species, S. invicta (sampling sites for this species are shown in Figure 1). A total of 164 unique sequences were identified, of which 121 were recovered from S. invicta. This is a large increase over the six alleles previously described from this species based on sampling at a single locality in Argentina and throughout the introduced USA range [10], [11]. The great majority of Gp-9 alleles in the complete fire ant data set (91%) were represented as singletons. Among the 13 alleles recovered from more than one specimen, three were shared between S. invicta and ants identified as belonging to the closely related species S. quinquecuspis or S. megergates (see Figure 2).

1	Comodoro
2	Pontes E Lacerda (5)
3	Pôrto Esperidião
4	Mirasol d'Oeste
5	São Pedra da Cipa
6	Rondonópolis
7	Pedra Preta (4)
8	Coxim (4)
9	Corumba
10	Miranda
11	São Gabriel do Oeste (4)
12	Campo Grande (4)
13	Nova Alvorada do Sul
14	Carapó
15	Guaira
16	Ceu Azul (7)
17	São Miguel d'Oeste
18	Clorinda (4)
19	Velaz
20	Formosa (17)
21	Resistencia
22	Barranqueras (5)
23	Corrientes (8)
24	Ita Ibate
25	Posadas
26	San Carlos (6)
27	Charada
28	Suncho Corral
29	Santiago del Estero (4)
30	Lajeado do Bugre
31	Soledade
32	Rinco dos Cabrais (4)
33	Curitiba
34	Rio Negro
35	Correia Pinto
36	Vacaria
37	Arroio dos Ratos (9)
38	Nova Petropolis
39	Pôrto Alegre
40	Las Toscas
41	San Justo
42	La Paz (6)
43	Coronda
44	Rosario (13)

Figure 1. Locations of sampling sites for S. invicta in South America. The native range of the species is indicated by gray shading on the map. The number of nests sampled at a site (= number of sequences obtained) is indicated in parentheses in the key to the sites if greater than one. Sites at which b-like alleles were found are highlighted with black rectangles; those at which the b alleles of the b-like class were found are underlined.

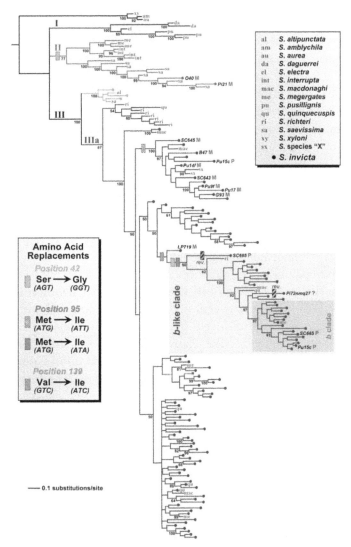

Figure 2. Summary hypothesis of phylogenetic relationships of fire ant Gp-9 sequences based on Bayesian inference. Sequences from species other than S. invicta are indicated by specific abbreviations in gray. Sequences from S. invicta are indicated by dots; the identity and social form (M, monogyne; P, polygyne; ?, unknown) of the colony of origin are also indicated for critical sequences. All colonies from which alleles of the b-like clade were recovered were confirmed to be polygyne. Inferred amino acid replacements at three codons considered to be diagnostic for b-like alleles are mapped onto the phylogeny (reversals are indicated by bars with cross-hatching). Posterior clade probabilities >50% are indicated as percentages below branches.

Plots of transition and transversion rates against pairwise sequence distances for various sets of alleles revealed little evidence of mutational saturation in the non-coding regions or at any of the three codon positions in exons (data not shown). For the entire data set, a consistent A+T bias in base composition was found, which is more substantial in the non-coding regions (0.816) than coding regions (0.574). No evidence of non-random codon usage was found over the entire set of sequences (scaled χ^2 = 0.104 using Yates correction; effective number of codons = 57.07 out of a maximum of 61; codon bias index = 0.348). Also, there is no evidence for intragenic recombination having occurred (all potential recombination events fall below 95th percentile significance level [DSS: 84%; LRT: 46%]). This latter result parallels the lack of evidence for recombination at Gp-9 reported by Krieger and Ross [11].

Phylogenetic Relationships of Gp-9 Alleles

All phylogenetic hypotheses recovered by the differing methods are compatible with one another. However, the Shimodaira and Hasegawa (SH) test identified the relatively poorly resolved maximum parsimony (MP) tree as significantly worse than the Bayesian inference (BI) tree (Δ-lnL = 149.233, P = 0.015), whereas the minimum evolution (ME) and BI trees did not differ significantly (Δ-lnL = 36.944, P = 0.376). The poor overall node resolution obtained with the MP analysis is consistent with the relatively few parsimony-informative sites (134) among the 164 unique sequences. The relatively high standard deviations of split frequencies in the BI analyses (~0.085) also suggest limited information content of our data set, which results in an inability to resolve some parts of the phylogeny with confidence (e.g., [23]).

The phylogenetic hypothesis produced by BI is shown in Figure 2. Three major Gp-9 allele clades of unresolved relationship to one another are apparent among the ingroup sequences (those recovered from the South American fire ants). Clade I contains sequences from three Solenopsis species considered to be rather distant relatives of S. invicta [24]. Clade II contains sequences from an assortment of species more or less closely related to S. invicta, as well as two S. invicta sequences. Clade III contains two relatively well supported lineages. One includes three S. invicta sequences along with sequences from two quite distant relatives (S. altipunctata and S. saevissima). The second, large lineage (Clade IIIa) contains only Gp-9 alleles recovered from the socially polymorphic South American fire ants (S. invicta and its close relatives S. richteri, S. megergates, S. quinquecuspis, S. macdonaghi, and S. species "X"). The relationships of the major Gp-9 lineages depicted in Figure 2 mirror the relationships inferred by Krieger and Ross [11] from a much smaller data set. Importantly, the previously detected paraphyly and polyphyly of S. invicta Gp-9 alleles with respect to those of the other socially polymorphic species [10], [11] are extended here to even more distantly related fire ant species.

Within the lineage comprising exclusively alleles from the socially polymorphic species (Clade IIIa), a clade composed almost entirely of sequences from S. richteri forms a sister lineage to the remaining sequences. This evidence for a basal position of S. richteri alleles within the group of alleles of all socially polymorphic species again is consistent with earlier conclusions based on smaller data sets [10], [11].

Significantly, the b-like allele clade is recovered once again in our study, confirming a monophyletic origin of the variants associated with polygyny in the South American fire ants. Support for such a clade varies depending on how it is defined. If defined only by possession of the Ile^{139} residue (the first of the three substitutions typical of b-like alleles to appear in the lineage), the clade is supported by a posterior probability value <0.5. However, if defined by possession of all three residues considered diagnostic for the b-like allele class, the posterior probability value increases to 0.5. (Note that the Val^{139} encoded by a colony SC665 sequence apparently represents a reversal; Figure 2). Given the evidence that sequences encoding all three residues are required for the expression of polygyny (below), we choose to define the b-like clade by the synapomorphy of joint possession of residues Gly^{42} and Ile^{95}, thus excluding from the group the sequence from colony LP719 encoding Ile^{139} but neither of these other residues (see Figure 2).

Finally, our data verify that the b alleles of S. invicta form a relatively recently derived monophyletic group within the b-like clade. This confirms earlier evidence that the radical, charge-changing Glu151Lys replacement characterizing these alleles occurred only once, presumably in an ancestral S. invicta population from northeastern Argentina or southeastern Brazil, where the alleles currently predominate (see Figure 1).

Nucleotide Variation at Gp-9

Most nucleotide sites (85%) are invariant across all Gp-9 sequences from the different Solenopsis species, and the two most divergent sequences (from S. aurea and S. invicta) differ at just 47 (2.1%) of their nucleotides. Most of the variable sites in the total collection of sequences (66%) occur in the non-coding regions. However, because these regions encompass 80% of the total sequence length, proportionately more variable sites occur in the coding regions (22%; 21% for third codon positions) than in the non-coding regions (14%).

The magnitude of variation at various positions along Gp-9 and its 3′ flanking region in all the study species is shown in Figure 3. This depiction of per-site nucleotide counts confirms that overall variation is not conspicuously lower in the exons than in the non-coding regions; indeed, the average number of different nucleotides per site is 1.23 in the exons (all sites as well as third codon positions) but only 1.14 in the non-coding regions. However, considerable heterogeneity exists within both types of regions. For instance, a notable spike in variation is evident in the 3' portion of exon 5, with 34% of nucleotide positions 1646-1708 (33% of the third codon positions) comprising variable sites (these sites correspond to amino acid residues 134-153). On the other hand, conspicuous dips in variation are apparent at nucleotide positions 620-715 in intron 2 (5% variable sites) and in the 3'-UTR (3'-untranslated region; 6% variable sites). Very similar patterns in the distribution of variation along Gp-9 were observed when just the sequences recovered from the socially polymorphic species or from S. invicta were considered (data not shown).

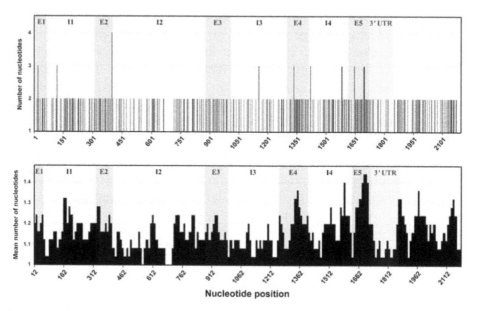

Figure 3. Nucleotide variation along Gp-9 and its 3' flanking region in all Solenopsis study species. The top graph shows the number of different nucleotides at each position; the bottom graph shows the mean number in a 25-bp sliding window moved by 10-bp increments. The darker shading indicates approximate positions of exons (E1-E5) and the light shading indicates the 3'-UTR. Introns are designated by I1-I4.

The great majority of unique alleles from S. invicta are highly similar to one another at the nucleotide sequence level; indeed, most alleles within the B-like and b-like classes differ by fewer than a dozen point substitutions (Figure 4). The two most divergent B-like alleles differ at only 36 of their nucleotides (1.5%), while the two most divergent b-like alleles differ at half that number. Considering both classes combined, an additional peak at 15–20 substitutions attributable to differences between B-like and b-like alleles becomes apparent (Figure 4).

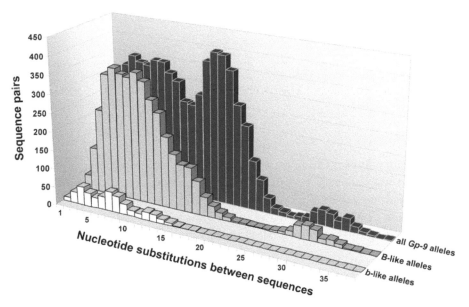

Figure 4. Distributions of numbers of nucleotide substitutions between pairs of unique Gp-9 sequences from S. invicta.

Several diversity statistics for Gp-9 from S. invicta are presented in Table 1. The mean number of substitutions between pairs of alleles (d) varies from less than one for third codon positions in b-like alleles to almost seven in non-coding regions when all alleles are combined. In parallel with the observed patterns of site variation across sequences from all the study species, the mean proportions of substitutions (p) between S. invicta coding-regions consistently exceed those between non-coding regions, regardless of whether all codon positions or just third positions are considered. While this difference between the coding and non-coding regions is marginally non-significant based on a Fisher's exact test (P = 0.071), a phylogeny-based resampling test showed that coding-region tree lengths always were considerably greater than any of the tree lengths derived from non-coding regions, under both MP and maximum likelihood (ML) criteria (thus, substitution rates in coding regions exceed those in non-coding regions at P<0.001). Considering only the b-like alleles of S. invicta, a greater than 3-fold excess of coding-region substitutions exists, attributable mostly to augmented variation in the first and second codon positions. Remarkably, given the phylogenetic restriction of the b-like clade and the fact that only 1/3 as many b-like as B-like S. invicta sequences were studied, values of both d and p over all codon positions are greater for the b-like group. These patterns hint at the possibility that positive selection has driven the molecular evolution of Gp-9 in the b-like clade.

Table 1. Diversity statistics for Gp-9 from S. invicta.

	d^a			p^b						
		Coding regions			Coding regions					
	Non-coding regions	All codon positions	Third codon positions	Non-coding regions	All codon positions	Third codon positions	Prop. variable codons	Prop. variable amino acids	Codon diversity	Amino acid diversity
b-like alleles	3.62	3.32	0.41	0.0021	0.0072	0.0027	0.1242	0.0980	0.0253	0.0221
B-like alleles	6.20	3.05	1.26	0.0036	0.0067	0.0082	0.3766	0.3052	0.0271	0.0211
all *Gp-9* alleles	6.66	5.41	1.76	0.0039	0.0118	0.0109	0.4351	0.3571	0.0362	0.0308

[a]Uncorrected mean number of nucleotide substitutions between all pairs of unique alleles [56]
[b]Uncorrected mean proportion of nucleotide substitutions between all pairs of unique alleles [56]

Apart from point substitutions, a single previously unknown structural change in Gp-9 also was detected. A sequence from an S. invicta colony in Santiago del Estero, Argentina, carries a unique point mutation in exon 5 that transforms the stop codon (TAA) at position 154 into a glutamine-encoding codon (CAA), thereby extending the C-terminal tail of the resulting protein by 22 amino acids.

Transition Bias at Gp-9

Ratios of the rates of transitions to those of transversions in the coding regions are shown for several data subsets in Table 2. Transition bias is negligible and similar between the third codon position and positions 1+2 for the B-like alleles. In contrast, a huge disparity in transition bias exists between these codon positions in the b-like alleles, due to the combined effects of a modest bias towards transitions at positions 1+2 and a sharp bias towards transversions at the third position. Over all studied sequences, there is a slight transition bias at the first two codon positions and a negligible bias at the third. These results suggest different patterns of selection acting on B-like and b-like alleles; because the great majority of third position transversions are nonsynonymous, their elevated rates only in the b-like clade are consistent with positive selection having acted specifically on this lineage (see [25]).

Table 2. Transition/transversion rate ratios (transition bias) for Gp-9 coding regions in Solenopsis.

		Codon position	
Class of *Gp-9* sequences		1+2	3
all *Solenopsis* sequences		3.01	1.30
Clade IIIa sequences	*b*-like	3.14	0.09
	B-like	1.23	0.81
S. invicta sequences	*b*-like	3.30	0.11
	B-like	1.07	1.22

Amino Acid Replacements at Gp-9

The sequence logo depicting the variable amino acids encoded by Gp-9 from all the study species (Figure 5) reveals that the great majority of polymorphic positions feature a single dominant residue along with a second minor residue represented in just one or two sequences. A very similar pattern is seen for sequences from just Clade IIIa or from just S. invicta, although several additional positions are monomorphic in each of these smaller data sets (data not shown). Major polymorphisms found in all the data sets at positions 39, 42, 117, 120, 151, and 152 feature substitutions between amino acids of different property groups. All of these positions except 151 also are implicated to be under positive selection by comparison of nonsynonymous and synonymous substitution rates (below). Position 42 is of special note because it features one of the three replacements defining the b-like allele clade.

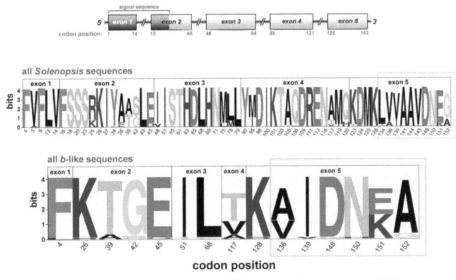

Figure 5. Sequence logos for variable amino acids encoded by unique Gp-9 sequences in all Solenopsis study species and in the b-like clade. Logos represent each position by a stack of letters, with the height of each letter proportional to the frequency of the amino acid in the recovered sequences. Overall stack height is proportional to the sequence conservation at that position (measured in bits, maximum sequence conservation is 4.3 bits). A schematic of the exon/intron structure of fire ant Gp-9 is shown above the logos. Codons in the highly variable 3' portion of exon 5 are demarcated with light grey boxes.

A surprisingly large number of amino acid replacements appear to have occurred during the evolution of the b-like clade, based on parsimony reconstruction. The three jointly diagnostic positions, 42, 95, and 139, feature replacements at or near the base of the clade (see Figure 2). Also, position 39 underwent one replacement at the base of and one within the clade, while position 117 underwent five replacements within the clade. Finally, positions 136, 151, and 152 underwent single replacements within the clade. Several of these replacements involve codons in the 3′ portion of exon 5 (Figure 5), thus paralleling the spike in nucleotide variation observed in this region (Figure 3).

Within S. invicta, over 43% of codons and 35% of amino acid residues are variable across all recovered alleles (Table 1), values that drop minimally for the B-like alleles but substantially for the b-like alleles. The similar values for the two metrics within each allele class suggest that the great majority of coding-region nucleotide substitutions have yielded amino acid replacements, a conclusion reinforced by the similarities between the codon diversity and amino acid diversity estimates. The prevalence of nonsynonymous substitutions is reflected also in the similar or higher estimates of p from all codon positions relative to third codon positions (Table 1). A similar prevalence of nonsynonymous coding-region substitutions exists across all the species; about 90% of codons exhibit identical codon and amino acid diversities, indicating that every observed nucleotide substitution at these locations caused an amino acid replacement.

An important finding of our survey is the existence of considerable amino acid variation across the three Gp-9 codons regarded as jointly diagnostic for the b-like and B-like allele classes in the socially polymorphic species. Alleles of the b-like class were reported earlier to always encode Gly^{42}, Ile^{95}, and Ile^{139} residues, whereas alleles of the B-like class generally were found to encode Ser^{42}, Met^{95}, and Val^{139} residues [10], [11]. We recovered 17 sequences from native S. invicta that feature some combination of b-like residues at one or two of these crucial positions and B-like residues at the remaining position(s). Two S. invicta sequences in Clade II (from colonies O^{40} and Pi^{21}) bear the b-like Gly^{42} residue together with the B-like Met^{95} and Val^{139} residues (Figure 2); remarkably, the Ser42Gly replacement characterizing this clade apparently occurred independently of the analogous replacement at the base of the b-like clade. A well supported group within Clade IIIa includes 12 S. invicta sequences encoding the b-like Ile^{95} residue together with the B-like Ser^{42} and Val^{139} residues; again, the Met95Ile replacement at the base of this group occurred independently of the analogous replacement in the stem lineage of the b-like clade (Figure 2). A sequence from colony LP719 encodes the b-like Ile139 residue together with the B-like Ser^{42} and Met^{95} residues. Finally, two S. invicta sequences within the b-like clade experienced apparent reversals from a b-like to B-like residue, at position 42 (colony Pi72nmq27) or position 139 (colony SC665) (Figure 2). The significance of these novel Gp-9 sequences with respect to the form of social organization expressed by the source colonies is explained next.

Association of Polygyny with b-Like Residues in S. Invicta

All 28 S. invicta colonies from which alleles encoding all three characteristic b-like residues were recovered were shown by microsatellite analysis to be polygyne (i.e., they contained multiple offspring matrilines). On the other hand, ten exemplar colonics yielding sequences representing all the major B-like clades were shown to be monogyne (i.e., they contained only a single offspring matriline). Thus, we further confirm an important conclusion from previous limited surveys in the native range [10], [26]; possession of typical b-like alleles by a colony's workers invariably is linked to the expression of polygyne social organization.

We also were able to determine the social form of 12 of the 17 S. invicta colonies that yielded novel Gp-9 variants encoding some combination of b-like and B-like residues at positions 42, 95, and 139. Data from the first ten such colonies listed in Table 3, all of which were monogyne, reveal that no single b-like residue is associated with polygyny. Significantly, the two polygyne colonies found to contain such novel variants (Pu15c, SC665) were found upon further sequencing of additional colony members to also contain workers with typical b-like alleles. Based on these data, we conclude that all three characteristic b-like residues may be jointly required for the expression of polygyny in the socially polymorphic South American fire ants.

Table 3. Association of colony social organization in S. invicta with amino acid residues at three Gp-9 codons jointly diagnostic for B-like and b-like alleles.

Colony	Colony social organization[a]	Codon[b]		
		42	95	139
O40	monogyne	*b*-like	B-like	B-like
Pi21	monogyne	*b*-like	B-like	B-like
SC645	monogyne	B-like	*b*-like	B-like
B47	monogyne	B-like	*b*-like	B-like
Pu14f	monogyne	B-like	*b*-like	B-like
SC643	monogyne	B-like	*b*-like	B-like
Pu9f	monogyne	B-like	*b*-like	B-like
Pu17	monogyne	B-like	*b*-like	B-like
G93	monogyne	B-like	*b*-like	B-like
LP719	monogyne	B-like	B-like	*b*-like
Pu15c[c]	**polygyne**	B-like	*b*-like	B-like
		b-like	*b*-like	*b*-like
SC665[c]	**polygyne**	*b-like*	*b-like*	B-like
		b-like	*b*-like	*b*-like

b-like residues and polygyne social organization are shown in bold for emphasis.

[a]Determined by microsatellite analysis

[b]B-like residues are Ser^{42}, Met^{95}, and Val^{139}, whereas b-like residues are Gly^{42}, Ile^{95}, and Ile^{139}

[c]Two different sequences were recovered from workers in these colonies; in each case, one represents a typical b-like allele (see Figure 2)

Selection on Gp-9

All of the selection analyses we undertook yielded evidence of selection of some form at various codons in Gp-9 and on various branches of the allele phylogeny. Among the site-specific methods, the single likelihood ancestor counting (SLAC) method identified two positively selected codons (Table 4), but the Suzuki-Gojobori counting method identified none. The former method also detected negative selection on codon 32, while both methods detected such selection on codon 99 (the two positions are invariant in the complete data set). The Bayesian random-effects method detected ten positively selected positions with very high confidence (posterior probability >95%) and another four with less confidence (posterior probability 90–95%) (Table 4). (A tendency of the counting methods to produce more conservative results than the random-effects method has been reported previously [27].) Three of the 14 positions implicated by the latter method as being positively selected (134, 145, and 152) are in the highly variable 3′ portion of exon 5 (Figures 3 and 5).

Table 4. Site-specific positive selection on Gp-9 in Solenopsis identified by different methods.

Codon	Method		
	Site-specific		Branch-site-specific
	SLAC counting[a] (*dN-dS, P*)	Bayesian random-effects (*dN/dS,* post. prob.[b])	Bayes empirical Bayes random-effects[c] (*dN/dS,* post. prob.)
39	2.49, 0.13	79, **1.00**	17.5, 0.957
42	–	79, **0.986**	–
45	–	79, **0.953**	–
48	–	79, **0.973**	–
61	–	79, 0.935	–
75	–	79, **0.995**	–
78	–	79, **0.998**	–
95	–	79, **0.993**	–
117	2.50, 0.13	79, **1.00**	17.5, 0.995
119	–	79, 0.900	–
120	–	79, **0.973**	–
134	–	79, **0.998**	–
145	–	79, 0.907	–
152	–	79, 0.908	–

Dashes indicate the absence of evidence for positive selection using a particular method.
[a] A nominal α-level of 0.25 is used to indicate statistical significance of selection for this method [27]
[b] Posterior probabilities >95% obtained using this method are shown in bold
[c] Only branches of the *b*-like clade were examined for positive selection using this method

Considering branch-specific selection, the counting method of Zhang et al. [28] revealed evidence of positive selection along three branches of the Gp-9 phylogeny. Two of these are the stem lineage and its succeeding descendant branch at the base of the b-like radiation (d_N/d_S = 3/0 [P = 0.004] and d_N/d_S = 2/0 [P = 0.026], respectively); significantly, the canonical b-like Ser42Gly and Met95Ile substitutions occur along the stem (see Figure 2). The lineage leading to the clade that includes the colony LP719 allele as well as the b-like alleles was not identified as experiencing selection (d_N/d_S = 1/0). A third branch under positive selection represents a relatively basal, well supported lineage within Clade IIIa consisting of one allele each from S. invicta and S. macdonaghi (d_N/d_S = 4/1 [P = 0.003]).

When we applied the branch-site-specific (Bayes empirical Bayes random-effects) method to the ancestral branch subtending the LP719 allele and b-like clade, d_N/d_S again did not differ significantly from one (d_N/d_S = 1.05, P = 0.75). On the other hand, the method did detect a ratio significantly greater than one over the stem and interior branches of the b-like clade (d_N/d_S = 17.5, P<0.00001). This approach identified with confidence two positions under positive selection on these b-like branches, 39 and 117, both of which also were identified by two of the site-specific methods applied over the entire tree (Table 4).

When we applied the branch-site-specific analysis to just the clade of b alleles (those b-like alleles bearing a charge-changing amino acid substitution at position 151), no significant signature of selection was detected on its stem lineage or internal branches.

Examination of overall dN-dS estimates for each pair of Gp-9 sequences from the socially polymorphic South American fire ants (Clade IIIa sequences) reveals a two-fold excess of positive values (Figure 6). Estimates obtained separately for the B-like and b-like classes show that the excess is especially marked for the latter class, consistent with evidence from several of the analyses above implicating positive selection in this clade of Gp-9 alleles associated with polygyny.

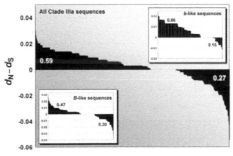

Figure 6. Distributions of overall pairwise estimates of d_N/d_S for Gp-9. Estimates are shown separately for the set of all Gp-9 sequences from the socially polymorphic South American fire ant species (Clade IIIa sequences) as well as for the subsets of all B-like and b-like sequences. Proportions of positive and negative dN-dS values are indicated for each group of sequences.

Phylogeography of Gp-9 Variants in S. Invicta

Our sequence data confirm the occurrence of b-like alleles at all five sampling sites in the south-central portion of the native range previously reported to contain such alleles (based on allele-specific PCR analyses [26]). In addition, we discovered alleles of this class at four other sites well outside of their previously known area of occurrence (Corumba, Coxim, La Paz, and Suncho Corral; see Figure 1). These results suggest that while polygyny seems to be concentrated in northern Argentina and southeastern Brazil, it may occur at some frequency through much of the native range of S. invicta.

An initial analysis of molecular variance (AMOVA) of B-like alleles revealed that no detectable variation occurs among the ten arbitrarily clustered groups of sites once the 30% of total variation found among sites within groups is accounted for (significance of among-site differentiation; P<0.001). Similar results were obtained in a second analysis of 14 sites clustered according to patterns of regional differentiation at 14 neutral nuclear genes; no variation occurs among the regional groups, but 21% of the total variation occurs among sites within groups (P = 0.001 for among-site differentiation). For comparison with the results of this second Gp-9 analysis, 14% of the variation at the neutral genes occurs among the regional groups, while 10% resides among sites within groups (P<0.001 for differentiation at both levels). Thus, Gp-9 in the monogyne form of native S. invicta appears to exhibit somewhat stronger differentiation than neutral nuclear markers at very local scales but much weaker differentiation at broader geographic scales covering hundreds of kilometers.

No significant pattern of isolation-by-distance was detected using Nei's DA values for the B-like alleles at 40 sites (one-tailed P = 0.087). Thus, differentiation in Gp-9 composition does not increase in parallel with geographic separation of sites. This finding again contrasts with the strong isolation-by-distance patterns detected using neutral nuclear markers in native S. invicta populations [29].

Discussion

The objective of this study was to survey naturally occurring molecular variation at Gp-9, a candidate gene of major effect on the expression of fire ant colony social organization, with a special emphasis on uncovering the extent and distribution of this variation in native populations of the well studied pest species Solenopsis invicta. Patterns of observed sequence variation at Gp-9 were examined with the following goals: i) to reconstruct the evolutionary relationships of variant sequences from S. invicta and its close relatives, ii) to identify important mutational factors affecting variation at the gene, iii) to examine the historical role of selection in shaping the variation, iv) to learn whether any single candidate amino acid residue in GP-9 protein is completely predictive of social organization, and v) to examine the geographical distribution of the observed variation. The motivation of the work was to help bridge the gap between functional biochemical information and molecular population genetic data by constructing a cogent evolutionary narrative of the genetic underpinnings of a major social adaptation (see [20]).

General Features of the Molecular Evolution of Gp-9

Our large survey of Gp-9 sequences from native fire ant populations succeeded in uncovering a large number of unique variants, with over 120 alleles found in S. invicta alone. Intragenic recombination was inferred to be unimportant in generating the diversity of sequences observed, a conclusion reached also by Krieger and Ross [11].

The exon/intron structure of all 164 fire ant Gp-9 variants is identical to that reported previously [10], [11], with one remarkable exception. A single S. invicta sequence contained a nonsynonymous nucleotide substitution in codon 154 that transformed it from a stop codon to a glutamine-encoding codon, thereby extending the C-terminal tail of the GP-9 protein by 22 amino acids. Two of these supernumerary residues are basic (Lys159, His175) while none is acidic, so that the mutant protein is likely to have a charge change mirroring or exceeding that of the proteins encoded by the distantly related b alleles of S. invicta (which have a unique Lys151 replacement). The charge change in the C-terminus of the b-encoded proteins is associated with recessive deleterious (lethal) effects not found in other b-like alleles [17]; these may stem from changes in the ligand binding/unloading properties or in the ability of the protein to form biologically active dimers, judging from the fact that the C-termini of odorant-binding proteins seem to be involved in these functions [30]. Demonstration of similar deleterious effects of the elongated mutant protein could pave the way for functional experiments intended to clarify some basic biochemical features of GP-9 protein.

Significant patterns in the nucleotide variation along Gp-9 and its 3' flanking region are evident when sequences from all the fire ant species are compared. First, the average amount of nucleotide variation in the exons exceeds that in the non-coding regions, based on the proportions of variable sites as well as the mean numbers of different nucleotides per site. Specifically, about one-fourth of coding-region nucleotide sites are variable but only half that proportion of non-coding sites are variable. The disparity persists regardless of whether all codon nucleotides or just third codon positions are considered, suggesting that most of the elevated exon nucleotide diversity consists of nonsynonymous substitutions. Superimposed on this general difference between coding and non-coding regions is considerable heterogeneity in the variation occurring within the two types of regions. Apparent elevation in variation above the background level for exons occurs in the 3' portion of exon 5 (34% vs. 22% variable sites), much of which translates into elevated amino acid variation. Moreover, apparent depressions in variation below the background non-coding level occur in the middle of intron 2 and in the 3'-UTR (<6% vs. 14% variable sites).

Very similar patterns of variation along Gp-9 are evident for just the sequences from the socially polymorphic species (Clade IIIa sequences) or from S. invicta. For S. invicta, the mean proportion of nucleotide substitutions (p) in the coding regions is elevated as much as 3-fold over that in non-coding regions, a difference judged to be highly statistically significant by a phylogeny-based resampling test (P<0.001). As is also true for the larger set of sequences, most coding-region nucleotide variation in S. invicta corresponds to amino acid replacements. With respect to absolute amounts of divergence over the entire gene region, only relatively modest nucleotide sequence differentiation occurs among the large number of Gp-9 alleles recovered from nominal S. invicta. Most pairs within the B-like or b-like allele classes differ by fewer than a dozen point substitutions across the 2300 bp sequence alignment, and the most divergent alleles from this species differ at just 38 (1.6%) of their sites. The two most divergent sequences in the entire data set (from S. aurea and S. invicta) differ at just 47 (2.1%) of their sites.

The resulting picture of Gp-9 sequence evolution in S. invicta and its fire ant relatives is that relatively few point mutations have accumulated across the gene, with only 15% of sites exhibiting variation, but a high proportion of these mutations occurred in the coding regions. Moreover, most of these coding-region substitutions led to amino acid replacements, so that in S. invicta alone over one-third of codons now encode variable residues. These findings raise several important points with respect to the molecular evolution of Gp-9. First, this pattern is consistent with a general lack of negative selection acting to constrain amino acid replacements over much of the encoded protein, as has been inferred also for other insect odorant-binding proteins based on their low amino acid sequence identities (the primary structure of these proteins evidently can be highly variable as long as the tertiary structure is conserved [18], [31]). Second, the elevated diversity observed in coding relative to non-coding regions may reflect different mutation rates or biases in the two types of sequence, which may in turn be related to the dramatically different base compositions [32]–[34]. Third, negative selection acting with greater overall force on the non-coding than coding regions may also play some role in the observed pattern of diversity; indeed, two non-coding tracts with extremely low variation across the surveyed sequences, including the entire

3'-UTR, potentially constitute evolutionarily constrained cis-regulatory elements [35], [36].

Finally, the elevated coding region nucleotide variation and high proportion of polymorphic amino acids might be construed as reflecting the historical action of positive selection, presumably an important general force in the evolution of insect odorant-binding proteins [37]. Direct evidence of some role for such selection on Gp-9 in fire ants comes from our formal selection analyses. Several codons were identified by various site-specific methods as having significantly elevated rates of nonsynonymous over synonymous substitutions across the species, and two of these (at positions 39 and 117) were implicated as well by a branch-site-specific method applied to the b-like clade. Moreover, replacements at the latter two codons typically involved residues of different property groups. Position 117 also was identified by Krieger and Ross [11] as subject to positive selection based on a more diverse set of Solenopsis Gp-9 sequences. Neither of these two consistently identified positions appears to be in the binding cavity or C-terminus of GP-9 protein based on earlier structure prediction analyses [11]. Thus, we cannot speculate about what physiological or other traits, if any, may be affected by the amino acid replacements at these locations.

More compelling evidence for positive selection having acted on Gp-9 comes from the combined results of the branch-specific and branch-site-specific methods, as well as the overall d_N-d_S estimates. These analyses consistently revealed elevated rates of amino acid replacement at the base of and throughout the b-like allele clade associated with polygyny in the socially polymorphic species. Similar results were obtained from the earlier selection analyses of Krieger and Ross [10], [11]. Congruent with these findings is our discovery of a highly elevated bias towards transversions at the third codon position in just the b-like lineage. The end result of this apparent burst of adaptive molecular evolution is a clade of rather similar alleles distinguished by low levels of silent substitutions but relatively high levels of amino acid replacements. Our study thus adds to the evidence that selection has played some creative role in the molecular evolution of Gp-9 in fire ants, primarily in the context of the origin and elaboration of an alternative form of social behavior.

It is worth emphasizing that adaptive divergence of Gp-9 generally does not seem to be associated with speciation events in the South American fire ants, a conclusion evident also from earlier studies [10], [11], [20]. This is inferred from the extensive paraphyly of Gp-9 sequences with respect to the nominal species as well as the persistence of the b-like clade as a trans-species polymorphism. Together with the evidence that selection has promoted the divergence of the b-like clade, these patterns imply that intraspecific social evolution is a more important driver of Gp-9 sequence diversification than is cladogenesis.

Two of the site-specific methods also yielded evidence of a single codon, at position 99, being under negative selection. This position is predicted to occur in the binding cavity of GP-9 [11], and so may represent a rare example where any variation in the encoded residue negatively impacts the binding capability and, hence, biological function of the protein.

Association of Gp-9 Molecular Variation with Social Organization

Among the coding-region diversity uncovered in our survey was substantial variation across the three codons that typically are jointly diagnostic between b-like and B-like alleles (positions 42, 95, 139) and, thus, predictive of colony social organization [10]. Rather than encoding alternate sets of amino acid residues at these three positions, the newly detected variants encode various combinations of b-like and B-like residues. Based on our determination of the social organization of the source colonies for these alleles, we conclude that no single residue is completely predictive of social behavior. Indeed, b-like residues invariably were present at all three positions in sequences from all polygyne colonies and, so, all of these residues may be jointly required for the expression of polygyny. Remarkably, placement of the novel variants in the Gp-9 allele phylogeny indicates that both the Gly^{42} and Ile^{95} b-like residues arose independently on at least two occasions. Apparently, however, only when they appeared concurrently on the stem lineage of the b-like clade already bearing the Ile^{139} b-like residue did expression of polygyny become possible.

Our proposal that all three characteristic b-like replacements are completely associated with polygyne behavior in the socially polymorphic species and, by implication, potentially involved in its expression, contradicts the conclusion of Krieger [30] and Krieger and Ross [11] that Val139Ile was the lone crucial replacement. This conclusion was based primarily on the inference from protein structure modeling that residue 139 forms part of the ligand-binding cavity. However, the same analysis also predicted that residue 95 lies in the binding cavity, and residue 42, although not expected to function in ligand binding, experienced a replacement between amino acids of different property groups. Although no study to date has identified codon 139 as being under positive selection, the Bayesian random-effects test implemented in this study implicated such selection on both codons 42 and 95. Moreover, our branch-specific analyses implicated positive selection on the stem lineage of the b-like clade, where the Ser42Gly and Met95Ile replacements occurred, but not on the preceding ancestral branch where the Val139Ile replacement occurred. We note that our data do not completely rule out the possibility that some combination of only two b-like residues may underlie polygyny, but testing this possibility must await discovery of colonies that contain such variants but lack typical b-like sequences.

Geographic Distribution of Gp-9 Molecular Variation

A previous survey of Gp-9 polymorphism in S. invicta using allele-specific PCR indicated that the b-like allele clade (and, by extension, polygyny) is restricted to the south-central portion of this species' range [26]. Our far more extensive sampling has extended the known area over which these alleles occur considerably northward into central Brazil. Nonetheless, in view of the fact that the ranges of the other socially polymorphic species harboring b-like alleles are restricted to eastern Argentina, Uruguay, and southeastern Brazil [38], an origin of the b-like clade in this area seems likely.

We found no indication of significant higher-level (regional) structure in the geographic distribution of B-like variants of Gp-9 in S. invicta, although local populations are highly differentiated from one another. This pattern stands in contrast to the striking regional differentiation observed for numerous other, presumably neutral, nuclear loci. One possible explanation for the difference is that, by chance, the distribution of Gp-9 variation does not closely track that of the remaining nuclear genome simply because of the probabilistic nature of allele lineage sorting (e.g., [39]). Another possibility is that sporadic interspecific hybridization in different areas followed by introgression of heterospecific Gp-9 alleles has broken down any geographic pattern that may have developed due to restricted inter-regional gene flow. Several lines of evidence support the plausibility of this scenario. First, the B-like alleles of S. invicta are extensively paraphyletic or polyphyletic with respect to the alleles of several other fire ants, including some species regarded as quite distant relatives [24]. Second, the S. invicta sequences that are polyphyletic with respect to these distant relatives often were obtained from colonies located within or close to the range of the other species (e.g., the closely related S. invicta and S. saevissima alleles in Clade II and the closely related S. invicta, S. saevissima, and S. altipunctata alleles in the sister lineage of Clade IIIA; see Figure 4). Finally, parallel patterns of minimal regional differentiation coupled with interspecific sequence paraphyly and polyphyly have been observed for the mtDNA of S. invicta [29], [40], with the latter features almost certainly the result of introgression.

This scenario posits that Gp-9 (and the mtDNA) flows more freely between fire ant species (and, perhaps, among regional S. invicta populations) than the bulk of the nuclear genome, perhaps because of a lack of selection against these introgressing elements (or, in the case of the mtDNA, because of selection favoring the spread of the cytoplasmic symbiont Wolbachia [41]). One important implication of this scenario, if true, is that polygyny may not have arisen in the common ancestor of the socially polymorphic fire ants, with the b-like lineage persisting through multiple speciation events [10], [11]. Rather, this allele class and the alternate form of social behavior with which it is associated conceivably arose more recently, then spread among species of the socially polymorphic clade through hybridization (see also [42]).

Conclusions

We recovered numerous Gp-9 alleles from S. invicta and other South American fire ants in their native ranges. Relatively little overall variation distiguishes these alleles, but the distribution of the variation along the gene and within the gene phylogeny is noteworthy. A surprising amount is found in the coding regions of the gene, with substitutions there usually causing amino acid replacements. Indeed, the proportion of variable amino acid positions is more than twice the proportion of variable nucleotide sites over the entire gene region, both across species and within S. invicta. The elevated coding-region variation may result from a general lack of negative selection acting to constrain amino acid replacements, different mutation rates or biases in coding and non-coding regions, negative selection acting with greater force on non-coding than coding regions, or most likely, positive selection acting on the protein in the b-like allele clade associated with polygyny. Finally, our determination of the social organization of key colonies confirmed the invariant link between the presence of typical b-like alleles and expression of polygyny, while our discovery of several novel alleles bearing various combinations of b-like and B-like codons revealed that no single amino acid residue is completely predictive of polygyne behavior.

This study thus yields information of use in bridging population genetic and functional approaches to understanding the genetic basis of polygyny in fire ants. With the inception of a broad, integrative approach to investigating the biochemical pathways in which the Gp-9 product functions, the phenotypic effects of molecular variation at Gp-9 and other pathway genes, and the potential involvement of other genes in linkage disequilibrium with Gp-9, substantial progress toward understanding the evolution of this key social adaptation can be expected.

Materials and Methods

Sampling

We obtained samples from several Solenopsis species of varying phylogenetic relationship to S. invicta [24] to assess the nature and extent of Gp-9 sequence variation in fire ants and to evaluate the monophyly of S. invicta sequences with respect to those of its closest relatives. These samples consisted of a single individual per nest collected in Argentina, Brazil, and the western USA between 1992 and 2001. Samples of the following species were obtained (numbers of individuals [nests] in parentheses): S. altipunctata (1), S. amblychila (1), S. aurea (1), S. daguerrei (2), S. electra (1), S. interrupta (4), S. macdonaghi (4), S. megergates (6), S. pusillignis (3), S. quinquecuspis (6), S. richteri (8), S. saevissima (9), S. xyloni (1), and the undescribed S. species "X" (2).

Samples of nominal S. invicta were obtained from 132 nests from 44 sites distributed over much of the native South American range (Figure 1) as well as from two sites in the introduced range in the USA (Georgia and California) between 1988 and 2004. Sites in the native range were chosen not only to maximize geographic coverage but also to include all of the genetically differentiated populations distinguished in earlier studies of neutral nuclear and mtDNA variation [29], [40], [43]. The purpose of this targeted sampling scheme was to uncover the maximal amount of diversity at Gp-9 in the focal species in order to generate a complete sequence phylogeny, which in turn formed the framework for our formal selection analyses. Multiple nests were sampled at many of the sites (see Figure 1), but generally only a single specimen was used from any single nest. The social organization of many of the sampled S. invicta colonies was determined previously by a combination of methods including discovery of multiple reproductive queens, determination of the number of offspring matrilines using allozyme markers, and detection of b-like Gp-9 alleles [26], [43], [44].

Live specimens of all the study species were collected directly from nests in the field then placed immediately on liquid nitrogen for transport back to the laboratory, where they were held in a −80°C freezer pending genetic analysis.

Sequencing of Gp-9

Sequencing methods followed the protocols of Krieger and Ross [10], [11], with some modifications. DNA was extracted using the Puregene DNA Isolation Kit (Gentra Systems, Minneapolis, MN). Polymerase chain reaction (PCR) reactions were set up in 10 µl volumes using 1.1× high fidelity PCR-ready reaction mix (Bio-X-Act Short Mix, Bioline, Randolph, MA) and 0.2 µM primers, with a hotstart thermal cycling regime starting at 95°C and followed by 35 cycles at 94°C (20 s), 62°C (30 s), and 68°C (1 min 40 s), and with a final elongation step at 68°C (10 min). Primer sequences were those used by Krieger and Ross [10] (Gp-9/-33 forward: 5′-CATTCAAAGTACAGTAGAATAACTGCC-3′, Gp-9_2218 reverse: 5′-CAGGAGTTTGAGTTTGTCACTGC-3′). The approximately 2200-bp amplification products included the full length 1700-bp Gp-9 gene (containing five exons and four introns) as well as a 500-bp segment of the 3′ flanking region containing the 170-bp UTR. These products were gel purified (QIAquick Gel Extraction Kit, Qiagen, Valencia, CA) and cloned into pCR2.1 vectors (Invitrogen, Carlsbad, CA), which were then used to transfect competent TOP10F′ E. coli cells (Invitrogen). Blue-white screening was used to identify positive clones, which were picked and subjected directly to a hotstart PCR amplification using M13 or the Gp-9 primers (same conditions as in the previous PCR but using Taq-Pro Complete, Denville Scientific Inc., Meutchen, NJ). The resulting PCR product was checked for correct length by running it out on an ethidium bromide-stained agarose gel, then it was purified using PEG 8000 (Promega, Madison, WI).

Methods for conducting DNA sequencing reactions using internal primers also followed the protocols of Krieger and Ross [10], [11]. Reactions were performed using the ABI PRISM BigDye Terminator v3.1 Cycle Sequencing Kit (Applied Biosystems, Foster City, Calif.), with the products run out in an ABI PRISM 3740xl DNA Sequencer (Applied Biosystems). In light of the considerable overlap of the internal sequencing reads, sequences were not determined in the reverse directions. Critical base calls in phylogenetically important sequences were confirmed by resequencing.

In order to ensure a sufficiently large sample of b-like alleles, clones derived from suspected polygyne S. invicta colonies were screened for such alleles using competitive allele-specific PCR [45]. Reaction mixes contained 0.13 μM primers, 0.33 μM complementary primer, 1× Taq-Pro Complete, and 0.5 μL of the clone PCR product; PCR was conducted using a cycling regime of 94°C (2 min), 35 cycles at 94°C (45 s), 64°C (45 s), and 72°C (1 min), and with a final elongation step at 72°C (5 min). The primers used in this allele-specific PCR recognize the single nucleotide substitutions at codons 95 and 139 considered to be jointly diagnostic of all b-like alleles [10], [21].

Determination of Colony Social Organization

Most colonies of unknown social organization yielding sequences that encoded one or more residues considered diagnostic for b-like alleles (Gly42, Ile95, Ile139) were subjected to microsatellite analyses to learn whether these colonies were monogyne or polygyne. Genotypes at eight loci (Sol-6, Sol-11, Sol-18, Sol-42, Sol-49, Sol-55, SolM-III, and SolM-V) were scored for 7–12 workers from each nest (PCR methods in [46]–[48]). PCR products were visualized using an ABI PRISM 3740xl DNA Sequencer. Queens of native S. invicta normally mate only once [44], [49], so the presence of more than three alleles at a locus among a colony's workers indicates the presence of multiple offspring matrilines (polygyny).

Genetic Analyses

All Gp-9 sequences were readily aligned by hand. The aligned sequences were tested for evidence of recombination using the DSS and LRT methods implemented in the program TOPALI [50].

Differentiation among Gp-9 alleles in their nucleotide composition was tested using homogeneity χ^2 analysis (implemented in the program PAUP* [51]) and visual inspection (implemented in the program SeqVis [52]). Non-random codon usage was tested using the program DNASP [53], with Yates' correction for the observed G+C content employed.

Transition and transversion rates were plotted against Tamura-Nei (TN93) distances for all pairs of sequences in various subsets of the data using the program DAMBE [54] to look for evidence of mutational saturation. In addition, the extent of transition bias (transition/transversion rate ratio) was estimated for the combined first and second codon positions and for the third codon position of exons in various data subsets using the program MEGA [55].

Several measures of nucleotide sequence variation were estimated for the Gp-9 alleles of nominal S. invicta. The uncorrected mean number of nucleotide substitutions (d) and proportion of nucleotide substitutions (p) were calculated for the coding regions (all positions and third codon positions) and the non-coding regions of all pairs of unique alleles [56] using MEGA. To measure the levels of observed nucleotide and amino acid variation in the coding regions of S. invicta sequences, we estimated two additional diversity indices for all unique sequences, the codon diversity and amino acid diversity [11]. Codon diversity denotes the variation at each in-frame coding-region triplet, while amino acid diversity denotes the variation at the corresponding amino acid residue. Values for the two measures range from zero to one for a given codon; a value of zero signifies a location with identical codons (or amino acids) in all sequences, whereas a value of one indicates that every unique sequence displays a unique codon (or amino acid) at this location. These indices were compared to determine the extent to which observed coding-region nucleotide variation translates into amino acid replacements. All of these analyses were performed separately for the b-like alleles, B-like alleles, and all alleles combined. Other measures of sequence variation that incorporate haplotype frequency estimates (such as π) were not estimated because our targeted sampling scheme would not yield unbiased frequency estimates.

We tested for differences in substitution rates between coding and non-coding regions of S. invicta sequences using two statistical approaches. First, a Fisher's exact test compared p between the two partitions. Second, a resampling test was used to compare tree lengths between coding and non-coding regions. A null distribution of tree lengths of non-coding nucleotides was generated by randomly resampling (without replacement) 462 such nucleotides from the original data matrix 1,000 times using MESQUITE [57], then calculating tree lengths for each resampled matrix on the Bayesian inference (BI) phylogeny under both maximum parsimony (MP) and maximum likelihood (ML) criteria using PAUP*. Tree lengths computed for the 462-bp coding region under these criteria on the same BI tree were compared to the tree lengths derived from the randomly resampled non-coding characters.

Analyses of Amino Acid Replacements

Patterns of amino acid variation at GP-9 were depicted graphically by generating sequence logos [58] using the program WEBLOGO [59]. Symbols were color-coded based on the chemical properties of each amino acid according to the scheme of Parry-Smith et al. [60]. Amino acid replacements were mapped onto the Gp-9 phylogeny using parsimony reconstruction as implemented in the program MACCLADE [61].

Phylogenetic Analyses

Sequences from the North American fire ant species S. amblychila, S. aurea, and S. xyloni were specified as outgroups for all phylogenetic analyses (see [24]). We assessed the potential impact on the phylogenetic analyses of heterogeneity in sequence nucleotide composition by constructing preliminary phylogenies with the Neighbor-Joining method [62], using the minimum evolution (ME) criterion based on either LogDet [63] or the ML distances between alleles (with all parameters estimated from the data). An SH test [64] conducted on the resulting two trees revealed no evidence of compositional heterogeneity (Δ-lnL = 7.543, P = 0.397).

Due to the prohibitive computational time required for even a single heuristic search under the MP criterion, we employed the parsimony ratchet method [65]; this approach was implemented by means of the program PAUPRAT [66] using 500 repetitions and randomly perturbing 25% of the characters for each re-weighting. The analysis was repeated ten times to ensure that tree space had been adequately searched, then repeated another ten times while considering gaps as character states.

Because the MP and ME phylogenies did not differ significantly according to an SH test (Δ-lnL = 112.3, P = 0. 124), we estimated the best fitting model of Gp-9 nucleotide evolution from the better resolved ME tree using the program MODELTEST [67]. We selected the most appropriate models for the complete data set and various partitions of it (non-coding regions; coding regions; first, second, and third codon positions) using the Akaike information criterion [68] and Bayesian information criterion [69] (see [70]).

Finally, we conducted four independent Markov chain Monte Carlo (MCMC) tree searches under the BI optimality criterion using the program MRBAYES [23], [71]. Multiple analyses were run to ensure adequate exploration of tree and parameter space [23], [72]. Five parallel chains, four of which were heated incrementally (temperature = 0.1), were started from random trees for each analysis, with initial parameter values based on the evolutionary model selected by MODELTEST. The chains were run for two million generations, with sampling every 100 generations. Stationarity of the chains was ascertained visually by plotting sample log-likelihoods through the course of each run, as well as by examining the convergence diagnostics (the potential scale reduction factor for all parameters approached 1.0 at stationarity) [23]. Pre-stationarity MCMC samples were discarded as burn-in (usually, around the first 700 samples), and the model parameters and tree topology were estimated using the remaining samples. The log-likelihoods, substitution models, and tree topologies were compared among independent runs using SH tests. After ensuring that all runs had converged on the same area in tree/parameter space, the samples from the four different runs were combined for final analysis.

Selection Analyses

Two general approaches for comparing nonsynonymous and synonymous substitution rates were employed to test for positive selection on Gp-9 in our complete data set, random-effects and counting analyses [27]. Fixed-effects analyses were not employed because they require a priori designation of sites evolving under different selective regimes [27]; we intended to use our data set for a largely independent test of the findings of Krieger and Ross [10], [11] and so wished to avoid biasing the results by focusing on specific sites previously identified as being under selection.

We employed two different random-effects methods, which fit a distribution of substitution rates across sites and then infer the rate at which each site evolves [73]. Because of computational limitations, the Bayesian method [74] was conducted on a reduced data set (25 exemplar sequences representing all major clades in the BI phylogeny) using the program MRBAYES. Coding and non-coding sites were separated into unlinked partitions for the analysis. Selection on the coding partition was estimated according to the M3 codon model [75], which is less restrictive than the commonly used Nielsen and Yang model [73] (see [23], [71]). MCMC chains were run for one million generations, with sampling every 100 generations. Otherwise, model specifications followed those used in the BI phylogeny estimation.

The second random-effects method used was a maximum likelihood branch-site-specific model [76], [77]. We removed all non-coding sites from the data set, then used the BI phylogeny as a guide to prune sequences that were sisters to sequences with identical coding regions or occurred in unresolved clades of identical coding sequences (a single exemplar was retained). This resulted in a reduced data set of 92 sequences. The CODEML program in the PAML package [78] was used to run the branch-site-specific model A, test 2 [77], which infers selection on codons along specified branches. We used the pruned BI phylogeny (with branch lengths estimated under the M3 model) to test for selection on the stem lineages and internal branches of the b-like allele clade as well as the b allele clade (the smaller clade of b-like alleles featuring the Glu151Lys substitution). We adopted the Bayes empirical Bayes [79] approach in place of the naive empirical Bayes approach to identify sites under selection [80]. The analysis incorporated the F3×4MG model of codon substitution [81] because it yielded better likelihood scores than the F3×4NY model [73]. We ran each analysis three times using different initial values for the parameters ω and κ.

We employed three different counting methods based on the Suzuki-Gojobori approach [82]. These methods estimate the number of nonsynonymous and synonymous substitutions at each codon position, then test for significant differences between the number of nonsynonymous changes per nonsynonymous site (dN) and number of synonymous changes per synonymous site (dS). The reduced data set of 92 sequences created for the branch-site-specific random-effects method was used also for the counting methods. The first counting method was the single likelihood ancestor counting (SLAC) analysis implemented in the program DATAMONKEY [83]. The global dN/dS ratio was estimated along with 95% confidence intervals using the HKY model of nucleotide substitution, with ambiguous characters resolved according to the most likely solution given by the model.

The second counting method we used follows more closely the original Suzuki and Gojobori [82] approach. This analysis was conducted in association with the branch-site-specific random-effects analysis using CODEML.

As a third counting method for detecting selection, we used the Zhang et al. method [28] adapted from the approach of Messier and Stewart [84]. This approach uses reconstructed ancestral sequences to test the null hypothesis of neutral evolution (dN = dS) along each branch of the inferred phylogeny by means of Fisher's exact tests. To maximize the statistical power of this test, we pooled non-coding sites with coding-region synonymous sites (e.g., [85]) after determining that there was no significant difference in substitution rates between these partitions (Fisher's exact test on p, P = 0.115). Ancestral sequences were reconstructed using the BASEML program in PAML, with dN and dS for each branch estimated using the "free-ratio" model [73]. We did not test for negative selection using this method.

Finally, we calculated maximum likelihood estimates of dN and dS for each pair of Gp-9 sequences from the socially polymorphic South American fire ants using CODEML to look for evidence of broad positive selection. These analyses also were conducted separately for the B-like and b-like alleles.

Tests such as the McDonald-Kreitman and HKA tests that compare intraspecific variation with interspecific divergence in order to detect signals of positive selection were not employed in this study because the major bouts of adaptive divergence in Gp-9 appear to be associated with social evolution within species rather than with speciation events. Also, tests using site-frequency spectrum data to test for deviations from neutral theory predictions (e.g., Tajima D test, Fay and Wu H test) were not employed because our targeted samples cannot provide the unbiased estimates of haplotype frequencies required for their proper application [86].

Phylogeographic Analyses

Evidence for geographical partitioning of related Gp-9 alleles in native S. invicta was examined by conducting a series of analysis of molecular variance (AMOVA) analyses [87] using the program ARLEQUIN [88]. This procedure partitions total genetic variation among the different sampling sites or clusters of sites in order to reveal hierarchical patterns of spatial differentiation. Only B-like alleles were included in order to avoid any effect of spatial restriction of polygyny (e.g., [26]). Genetic distances between Gp-9 alleles were estimated as the squares of the numbers of pairwise sequence differences. In an initial analysis, all 40 sites containing at least one B-like allele were clustered arbitrarily into ten regional groups (these groups occupied areas 100–300 km in diameter). In a second analysis, only the 14 sites for which three or more sequences were available were used. In this case, goups of sites were clustered on the basis of patterns of regional differentiation previously detected at 14 presumed neutral nuclear loci [29] (these groups occupied areas 100–600 km in diameter). Parallel AMOVA analyses using the neutral nuclear data of Ross et al. [29] from the same 14 sites were conducted to provide a direct comparison with the Gp-9 results from the second analysis. All alleles at each neutral locus were assumed to be equally related to one another (i.e., an infinite alleles model of mutation was assumed). Statistical significance of genetic differentiation among sites or clusters of sites was determined by permuting alleles across individuals (20,000 replicates) for Gp-9 or bootstrapping over loci (10,000 replicates) for the neutral markers.

Isolation-by-distance analyses were conducted for the Gp-9 sequences of S. invicta to learn whether differentiation between sites in their allele composition increases in parallel with their geographic separation. Only the 40 sites from which one or more B-like alleles were sampled were considered. The program GENE-POP [89] was used to examine the relationship of Nei's net number of nucleotide differences (DA [90]) with the natural logarithm of geographic distances between sites (see [91], [92]). Significance of isolation-by-distance relationships was determined by means of Mantel tests based on 10,000 data permutations coupled with estimation of Spearman rank correlation coefficients.

Authors' Contributions

Conceived and designed the experiments: KR DG. Performed the experiments: DG DS. Analyzed the data: KR DG. Contributed reagents/materials/analysis tools: KR DG DS. Wrote the paper: KR DG DS.

References

1. Gillespie JH (1991) The causes of molecular evolution. Oxford: Oxford University Press.
2. Hedrick PW (2004) Genetics of populations, 3rd edition. Boston: Jones and Bartlett.
3. Golding GB, Dean AM (1998) The structural basis of molecular adaptation. Mol Biol Evol 15: 355–369.
4. Wheat CW, Watt WB, Pollock DD, Schulte PM (2006) From DNA to fitness differences: sequences and structures of adaptive variants of Colias phosphoglucose isomerase (PGI). Mol Biol Evol 23: 499–512.
5. Nachman MW (2005) The genetic basis of adaptation: lessons from concealing coloration in pocket mice. Genetica 123: 125–136.

6. Phillips PC (2005) Testing hypotheses regarding the genetics of adaptation. Genetica 123: 15–24.

7. Vasemägi A, Primmer CR (2005) Challenges for identifying functionally important genetic variation: the promise of combining complementary research strategies. Mol Ecol 14: 3623–3642.

8. Ross KG (1997) Multilocus evolution in fire ants: effects of selection, gene flow, and recombination. Genetics 145: 961–974.

9. Ross KG, Keller L (1998) Genetic control of social organization in an ant. Proc Natl Acad Sci USA 95: 14232–14237.

10. Krieger MJB, Ross KG (2002) Identification of a major gene regulating complex social behavior. Science 295: 328–332.

11. Krieger MJB, Ross KG (2005) Molecular evolutionary analyses of the odorant-binding protein gene Gp-9 in fire ants and other Solenopsis species. Mol Biol Evol 22: 2090–2103.

12. Ross KG, Keller L (2002) Experimental conversion of colony social organization by manipulation of worker genotype composition in fire ants (Solenopsis invicta). Behav Ecol Sociobiol 51: 287–295.

13. Gotzek D, Ross KG (2007) Experimental conversion of colony social organization in fire ants (Solenopsis invicta): worker genotype manipulation in the absence of queen effects. J Ins Behav. In press.

14. Bourke AFG, Franks NR (1995) Social evolution in ants. Princeton: Princeton University Press.

15. Ross KG, Keller L (1995) Ecology and evolution of social organization: insights from fire ants and other highly eusocial insects. Annu Rev Ecol Syst 26: 631–656.

16. Tschinkel WR (2006) The fire ants. Cambridge, Massachusetts: Harvard University Press.

17. Hallar BL, Krieger MJB, Ross KG (2007) Potential cause of lethality of an allele implicated in social evolution in fire ants. Genetica 131: 69–79.

18. Vogt RG (2005) Molecular basis of pheromone detection in insects. In: Gilbert L, Iatrou K, Gill S, editors. Endocrinology, vol. 3. Comprehensive molecular insect science. London: Elsevier. pp. 753–804.

19. Keller L, Ross KG (1998) Selfish genes: a green beard in the red fire ant. Nature 394: 573–575.

20. Gotzek D, Ross KG (2007) Genetic regulation of colony social organization in fire ants: an integrative overview. Q Rev Biol 82: 201–226.

21. Ross KG, Krieger MJB, Shoemaker DD (2003) Alternative genetic foundations for a key social polymorphism in fire ants. Genetics 165: 1853–1867.

22. Eyre-Walker A (2006) The genomic rate of adaptive evolution. Trends Ecol Evol 21: 569–575.

23. Ronquist F, Huelsenbeck JP, van der Mark P (2005) MrBayes 3.1 manual. Available: http://mrbayes.csit.fsu.edu/mb3.1_manual.pdf. Accessed 11 November 2006.

24. Pitts JP, McHugh JV, Ross KG (2005) Cladistic analysis of the fire ants of the Solenopsis saevissima species-group (Hymenoptera: Formicidae). Zool Scripta 34: 493–505.

25. Bofkin L, Goldman N (2007) Variation in evolutionary processes at different codon positions. Mol Biol Evol 24: 513–521.

26. Mescher MC, Ross KG, Shoemaker DD, Keller L, Krieger MJB (2003) Distribution of the two social forms of the fire ant Solenopsis invicta (Hymenoptera: Formicidae) in the native South American range. Ann Ent Soc Amer 96: 810–817.

27. Kosakovsky Pond SL, Frost SWD (2005) Not so different after all: a comparison of methods for detecting amino acid sites under selection. Mol Biol Evol 22: 1208–1222.

28. Zhang J, Kumar S, Nei M (1997) Small-sample tests of episodic adaptive evolution: a case study of primate lysozymes. Mol Biol Evol 14: 1335–1338.

29. Ross KG, Krieger MJB, Keller L, Shoemaker DD (2007) Genetic variation and structure in native populations of the fire ant Solenopsis invicta: evolutionary and demographic implications. Biol J Linn Soc. volume 92: 541–560.

30. Krieger MJB (2005) To b or not to b: a pheromone-binding protein regulates colony social organization in fire ants. Bioessays 27: 91–99.

31. Nagnan-Le Meillour P, Jacquin-Joly E (2003) Biochemistry and diversity of insect odorant-binding proteins. In: Blomquist GJ, Vogt RG, editors. Insect pheromone biochemistry and molecular biology: The biosynthesis and detection of pheromones and plant volatiles. London: Elsevier. pp. 509–537.

32. Filipski J (1988) Why the rate of silent codon substitutions is variable within a vertebrate's genome. J Theor Biol 134: 159–164.

33. Haddrill PR, Charlesworth B, Halligan DL, Andolfatto P (2005) Patterns of intron sequence evolution in Drosophila are dependent upon length and GC content. Genome Biol 6: R67.

34. Ko W-Y, Piao S, Akashi H (2006) Strong regional heterogeneity in base composition evolution on the Drosophila X chromosome. Genetics 174: 349–362.

35. Wagner GP, Fried C, Prohaska SJ, Stadler PF (2004) Divergence of conserved non-coding sequences: rate estimates and relative rate tests. Mol Biol Evol 21: 2116–2121.

36. Andolfatto P (2005) Adaptive evolution of non-coding DNA in Drosophila. Nature 437: 1149–1152.

37. Forêt S, Maleszka R (2006) Function and evolution of a gene family encoding odorant binding-like proteins in a social insect, the honey bee (Apis mellifera). Genome Res 16: 1404–1413.

38. Pitts JP, McHugh JV, Ross KG (2007) Revision of the fire ants of the Solenopsis saevissima species-group (Hymenoptera: Formicidae). Zootaxa. In press.

39. Rosenberg NA (2003) The shapes of neutral gene genealogies in two species: probabilities of monophyly, paraphyly, and polyphyly in a coalescent model. Evolution 57: 1465–1477.

40. Shoemaker DD, Ahrens ME, Ross KG (2006) Molecular phylogeny of fire ants of the Solenopsis saevissima species-group based on mtDNA sequences. Mol Phyl Evol 38: 200–215.

41. Ahrens M, Shoemaker DD (2005) Evolutionary history of Wolbachia infections in the fire ant Solenopsis invicta. BMC Evol Biol 5: 35.

42. Keller L, Parker JD (2002) Behavioral genetics: a gene for supersociality. Curr Biol 12: 180–181.

43. Ross KG, Shoemaker DD (2005) Species delimitation in native South American fire ants. Mol Ecol 14: 3419–3438.

44. Ross KG, Krieger MJB, Shoemaker DD, Vargo EL, Keller L (1997) Hierarchical analysis of genetic structure in native fire ant populations: results from three classes of molecular markers. Genetics 147: 643–655.

45. Imyanitov EN, Buslov KG, Suspitsin EN, Kuligina ES, Belogubova EV, et al. (2002) Improved reliability of allele-specific PCR. Biotechniques 33: 484–490.

46. Krieger MJB, Keller L (1997) Polymorphism at dinucleotide microsatellite loci in fire ant Solenopsis invicta populations. Mol Ecol 6: 997–999.

47. Chen YP, Lu LY, Skow LC, Vinson SB (2003) Relatedness among co-existing queens within polygyne colonies of a Texas population of the fire ant, Solenopsis invicta. Southwest Ent 28: 27–36.

48. Shoemaker DD, DeHeer CJ, Krieger MJB, Ross KG (2006) Population genetics of the invasive fire ant Solenopsis invicta (Hymenoptera: Formicidae) in the United States. Ann Ent Soc Amer 99: 1213–1233.

49. Ross KG, Vargo EL, Keller L, Trager JC (1993) Effect of a founder event on variation in the genetic sex-determining system of the fire ant Solenopsis invicta. Genetics 135: 843–854.

50. Milne I, Wright F, Rowe G, Marshal DF, Husmeier D, et al. (2004) TOPALi: software for automatic identification of recombinant sequences within DNA multiple alignments. Bioinformatics 20: 1806–1807.

51. Swofford DL (2003) PAUP*; Phylogenetic analysis using parsimony (*and other methods). Version 4. Sunderland, Massachusetts: Sinauer.

52. Ho JWK, Adams CE, Lew JB, Matthews TJ, Ng CC, et al. (2006) SeqVis: Visualization of compositional heterogeneity in large alignments of nucleotides. Bioinformatics 22: 2162–2163.

53. Rozas J, Sanchez-DelBarrio JC, Messeguer X, Rozas R (2003) DNASP, DNA polymorphism analyses by the coalescent and other methods. Bioinformatics 19: 2496–2497.

54. Xia X, Xie Z (2001) DAMBE: software package for data analysis in molecular biology and evolution. J Hered 92: 371–373.

55. Kumar S, Tamura K, Nei M (2004) MEGA 3: Integrated software for molecular evolutionary genetics analysis and sequence alignment. Brief Bioinf 5: 150–163.

56. Nei M, Kumar S (2000) Molecular evolution and phylogenetics. New York: Oxford University Press.

57. Maddison WP, Maddison DR (2006) MESQUITE: a modular system for evolutionary analysis. Available: http://mesquiteproject.org. Accessed 4 September 2007.

58. Schneider TD, Stephens RM (1990) Sequence logos: a new way to display consensus sequences. Nuc Acids Res 18: 6097–6100.

59. Crooks GE, Hon G, Chandonia JM, Brenner SE (2004) WebLogo: A sequence logo generator. Genome Res 14: 1188–1190.

60. Parry-Smith DJ, Payne AWR, Michie AD, Attwood TK (1998) CINEMA– A novel color interactive editor for multiple alignments. Gene 221: GC57– GC63.

61. Maddison DR, Maddison WP (2000) MacClade 4: Analysis of phylogeny and character evolution, CD-ROM ed. Sunderland, Massachusetts: Sinauer.

62. Saitou N, Nei M (1987) The neighbor-joining method; a new method for reconstructing phylogenetic trees. Mol Biol Evol 4: 406–425.

63. Lockhart PJ, Steel MA, Hendy MD, Penny D (1994) Recovering evolutionary trees under a more realistic model of sequence evolution. Mol Biol Evol 11: 605–612.

64. Shimodaira H, Hasegawa M (1999) Multiple comparisons of log-likelihoods with applications to phylogenetic inference. Mol Biol Evol 16: 1114–1116.

65. Nixon KC (1999) The Parsimony Ratchet, a new method for rapid parsimony analysis. Cladistics 15: 407–414.

66. Sikes DS, Lewis PO (2001) Software manual for PAUPRat: A tool to implement parsimony ratchet searches using PAUP*. Available: http://www.ucalgary.ca/dsikes/software2.htm. Accessed 20 December 2006.

67. Posada D, Crandall KA (1998) MODELTEST: testing the model of DNA substitution. Bioinformatics 14: 817–818.

68. Akaike H (1974) New look at statistical model identification. IEEE Trans Auto Contr 19: 716–723.

69. Schwarz G (1978) Estimating the dimension of a model. Ann Stat 6: 461–464.

70. Posada D, Buckley TR (2004) Model selection and model averaging in phylogenetics: advantages of Aikake information criterion and Bayesian approaches over likelihood ratio tests. Syst Biol 53: 793–808.

71. Huelsenbeck JP, Ronquist F (2001) MRBAYES: Bayesian inference of phylogenetic trees. Bioinformatics 17: 754–755.

72. Larget B (2005) Introduction to Markov chain Monte Carlo methods in molecular evolution. In: Nielsen R, editor. Statistical methods in molecular evolution. New York: Springer.

73. Nielsen R, Yang Z (1998) Likelihood models for detecting postively selected amino acid sites and applications to the HIV-1 envelope genes. Genetics 148: 929–936.

74. Huelsenbeck JP, Dyer KA (2004) Bayesian estimation of positively selected sites. J Mol Evol 58: 661–672.

75. Yang Z, Nielsen R, Goldman N, Krabbe Pedersen A-M (2000) Codon-substitution models for heterogeneous selection pressure at amino acid sites. Genetics 155: 431–449.

76. Yang Z, Nielsen R (2002) Codon-substitution models for detecting molecular adaptation at individual sites along specified lineages. Mol Biol Evol 19: 908–917.

77. Zhang J, Nielsen R, Yang Z (2005) Evaluation of an improved branch-site likelihood method for detecting positive selection at the molecular level. Mol Biol Evol 22: 2472–2479.

78. Yang Z (1997) PAML: a program package for phylogenetic analysis by maximum likelihood. Appl Biosci 13: 555–556.

79. Yang Z, Wong WSW, Nielsen R (2005) Bayes empirical bayes inference of amino acid sites under positive selection. Mol Biol Evol 22: 1107–1118.

80. Scheffler K, Seoighe C (2005) A Bayesian model comparison approach to inferring positive selection. Mol Biol Evol 22: 2531–2540.

81. Muse SV, Gaut BS (1994) A likelihood approach for comparing synonymous and nonsynonymous nucleotide substitution rates, with application to the chloroplast genome. Mol Biol Evol 11: 715–724.

82. Suzuki Y, Gojobori T (1999) A method for detecting positive selection at single amino acid sites. Mol Biol Evol 16: 1315–1328.

83. Kosakovsky Pond SL, Frost SWD (2005) DATAMONKEY: rapid detection of selective pressure on individual sites of codon alignments. Bioinformatics 21: 2531–2533.

84. Messier W, Stewart C-B (1997) Episodic adaptive evolution of primate lysozymes. Nature 385: 151–154.

85. Rooney AP, Zhang J (1999) Rapid evolution of a primate sperm protein: relaxation of functional constraint or positive Darwinian selection? Mol Biol Evol 16: 706–710.

86. Yang Z (2006) Computational molecular evolution. New York: Oxford University Press.

87. Excoffier L, Smouse PE, Quattro JM (1992) Analysis of molecular variance inferred from metric distances among DNA haplotypes: application to human mitochondrial DNA restriction data. Genetics 131: 479–491.

88. Excoffier L, Laval G, Schneider S (2005) Arlequin ver. 3.0: an integrated software package for population genetics data analysis. Evol Bioinf Online 1: 47–50.

89. Raymond M, Rousset F (1995) GENEPOP (version 1.2): population genetics software for exact tests and ecumenicism. J Hered 86: 248–249.

90. Nei M, Li WH (1979) Mathematical model for studying genetic variation in terms of restriction endonucleases. Proc Natl Acad Sci USA 76: 5269–5273.

91. Slatkin M (1993) Isolation by distance in equilibrium and non-equilibrium populations. Evolution 47: 264–279.

92. Rousset F (1997) Genetic differentiation and estimation of gene flow from F-statistics under isolation by distance. Genetics 145: 1219–1228.

CITATION

Gotzek D, Shoemaker DD, and Ross KG. Molecular Variation at a Candidate Gene Implicated in the Regulation of Fire Ant Social Behavior. PLoS ONE 2(11): e1088. doi:10.1371/journal.pone.0001088. Originally published under the Creative Commons Attribution License, http://creativecommons.org/licenses/by/3.0/

Ultrasonic Communication in Rats: Can Playback of 50-kHz Calls Induce Approach Behavior?

Markus Wöhr and Rainer K. W. Schwarting

ABSTRACT

Rats emit distinct types of ultrasonic vocalizations, which differ depending on age, the subject's current state and environmental factors. Since it was shown that 50-kHz calls can serve as indices of the animal's positive subjective state, they have received increasing experimental attention, and have successfully been used to study neurobiological mechanisms of positive affect. However, it is likely that such calls do not only reflect a positive affective state, but that they also serve a communicative purpose. Actually, rats emit the highest rates of 50-kHz calls typically during social interactions, like reproductive behavior, juvenile play and tickling. Furthermore, it was recently shown that rats emit 50-kHz calls after separation from conspecifics. The aim of the present study was to test the communicative value of such 50-kHz calls. In a first ex-periment, conducted in juvenile rats situated singly on a radial maze appara-tus, we showed that 50-kHz calls can induce behavioral activation and ap-proach responses, which were selective to 50-kHz signals, since presentation of 22-kHz calls, considered to be aversive or threat signals, led to behavioral inhibition. In two other experiments, we used either natural 50-kHz calls, which had been previously recorded from other rats, or artificial sine wave stimuli, which were identical to these calls with respect to peak frequency, call length and temporal appearance. These signals were presented to either juve-nile (Exp. 2) or adult (Exp. 3) male rats. Our data clearly show that 50-kHz signals can induce approach behavior, an effect, which was more pronounced in juvenile rats and which was not selective to natural calls, especially in adult rats. The recipient rats also emitted some 50-kHz calls in response to call presentation, but this effect was observed only in adult subjects. Together, our data show that 50-kHz calls can serve communicative purposes, name-ly as a social signal, which increases the likelihood of approach in the recipi-ent conspecific.

Introduction

Rats emit distinct types of ultrasonic vocalizations (USV), which differ depending on age, the subject's current state and environmental factors [1]–[3]. Rat pups typically exhibit USV in response to isolation from mother and litter [4]. Juvenile and adult rats, on the other hand, produce two different types of USV, which have been classified primarily on the basis of their sound frequency as low and high frequency vocalizations.

Low frequency vocalizations, often termed 22-kHz calls, are emitted when rats are exposed to predators [5], foot-shocks [6]–[10], during inter-male aggression [11], [12], drug withdrawal [13], [14], handling [15], and social isolation [16]. Remarkably, anxiolytic drugs can reduce such vocalizations [17]–[19]. Function-ally, it was assumed that 22-kHz calls reflect a negative affective state akin anxiety and sadness [8], [9], and that they serve as alarm cries [5].

Conversely, high-frequency vocalizations, often termed 50-kHz calls, occur during or in anticipation of juvenile rough-and-tumble play [19], [20], mating [21]–[28], food consumption [29], electrical self-stimulation of the brain [29], [30], and addictive drugs [31]–[35]. Furthermore, rats also emit such calls when tickled by a skilled experimenter in a playful way [36]–[40], and rates of 50-kHz calls were found to be positively correlated with the rewarding value of tickle stimulation as measured by instrumental approach behavior [36], [37], [39].

Conversely, aversive stimuli including bright light [20], [37], predatory odors [37], the presence of foot shock cues [29] and drugs with aversive properties decrease levels of 50-kHz calls [41]. Based on such evidence, Panksepp and Burgdorf [40] suggested that 50-kHz calls might provide an archaic form of human laughter ("rat laughter"), which might serve as an index of the animal's subjective state [2]. Thereby, 50-kHz calls might provide a new and unique measure for analyzing natural reward circuits in the brain [29], [30], [42].

Recently, however, it was shown that 50-kHz calls can also occur in situations that are not necessarily pleasurable or even mildly aversive to rats. Thus, it was found that 50-kHz calls were emitted during short social isolation in the animal's own, or in a new soiled or fresh housing cage, irrespective of whether the animal's motivational status was high or low, i.e. irrespective of whether the animal was food-deprived or fed ad libitum [40], [43]. Also, during testing in an open field and an elevated plus maze 50-kHz calling was observed [43]. These findings are in line with observations of 50-kHz calls in various experimental controls, like naïve rats that were placed into a test arena containing fresh bedding [24], [44], or saline-injected rats in drug studies [33]–[35], [41]. Remarkably, the propensity to call differed dependent on the time-point of the last social contact, i.e. rats emitted 50-kHz calls primarily initially after separation from the cage mate [43]. Finally, it was found that not only the animal, which was isolated in a new housing cage emitted 50-kHz calls, but also the cage mate that remained alone in the home cage after the removal of the test rat [43]. These findings corroborated the idea that 50-kHz calls serve for communicative purposes, e.g. to (re)establish or keep contact.

A social function of rat USV was already confirmed successfully by performing playback studies in pups [45]–[47]. In adult rats, it was shown that the presentation of natural 22-kHz calls or 20-kHz sine wave tones can activate the fight/flight/freeze system [48]–[53]. However, little is known about the effects of 50-kHz calls on the behavior of the receiver. Schleidt [54] found that diverse artificial ultrasonic stimuli elicit Preyer's reflex, i.e. twitches of the auricles, in rats, and Thomas et al. [55] observed a suppression of instrumental bar pressing and bradycardia when artificial 50-kHz tones were presented. Apart from these early studies, responses to playback of high-frequency ultrasonic stimuli have been studied primarily within the sexual context. Here, changes in approach behavior [56], [57], proceptive behavior [22], [25], [27] and ultrasonic calling were observed [58]. Finally, two recent studies in non-sexual contexts obtained incongruent results. Burgdorf et al. [32] found that rats show instrumental behavior to receive playback of 50-kHz calls, whereas Endres et al. [59] did not find overt behavioral effects of 50-kHz playback.

The aim of the present study was to test the communicative value of 50-kHz calls by measuring overt and calling behavior during playback of such calls. As a testing environment, we used an unbaited radial-arm maze, since this apparatus had proven its usefulness in a previous experiment, where we had tested the behavioral effects of presenting pup 40-kHz calls to rat dams [47]. Here, it was hypothesized that presentations of 50-kHz calls induce locomotor activity and ultrasonic calling, whereas 22-kHz calls induce locomotor inhibition and a reduction in ultrasonic calling (Exp. 1). Furthermore, it was hypothesized that the 50-kHz call induced activation is stimulus-directed, i.e. that animals will approach the source of 50-kHz calls while calling themselves. Also, we assumed that the behavioral response is dependent on subject- and call-related features. Regarding subjects, we used juvenile (Exp. 1 & 2) and adult rats (Exp. 3), expecting stronger behavioral responses in juvenile rats, where 50-kHz calls occur in great numbers [37]. To test the effect of call features, natural 50-kHz calls and artificial sine wave tones (i.e. "calls" without amplitude and frequency modulation) were used (Exp. 2 & 3). In accordance to a bulk of evidence showing that primarily frequency modulated 50-kHz calls are linked to a positive affective state [30], [32], [42], it was expected that they can induce approach behavior. However, it was expected that flat 50-kHz signals might also induce approach behavior, since it was shown that flat calls are predominantly emitted after separation from the cage mate, suggesting that this call serves as a contact call [40], [43].

Materials and Methods

Animals and Housing

In total, 68 male Wistar rats (HsdCpb:WU, Harlan-Winkelmann, Borchen, Germany) served as subjects. In Exp. 1, 12 juvenile male rats were used, weighing 66.7±2.5 g (range: 52.5–76.5 g; about 25 days of age) on the test day. Twenty juvenile male rats were used in Exp. 2, weighing 80.9±1.5 g (range: 66.0–91.0 g; about 27 days of age) on the test day. Finally, 36 adult male rats were used in Exp. 3, weighing 320.5±6.3 g (range: 273.0–422.0 g; about 12 weeks of age) on the test day. All animals were naïve, except for animals of Exp. 2, which were separated from mother and litter two times for 10 min on postnatal day 11. Animals were housed in groups of 5 (Exp. 2) or 6 (Exp. 1 & 3) on Tapvei peeled aspen bedding (indulab ag, Gams, Switzerland) in a Macrolon type IV cage (size: 378×217×180 mm, plus high stainless steel covers). Lab chow (Altromin, Lage, Germany) and water (0.0004% HCl-solution) were available ad libitum. Animals were housed in an animal room with a 12:12 h light/dark cycle (lights on 7–19 h) where the environmental temperature was maintained between 20–25° Celsius.

Prior to testing, all animals were handled for 3 days in a standardized way (5 min each day).

Experimental Setting

Testing was performed on a radial maze of gray plastic with 8 arms (9.8×40.5 cm) extending radially from a central platform (diameter: 24 cm), which was elevated 52 cm above the floor (for details see: [60]). Acoustic stimuli were presented through an ultrasonic speaker (ScanSpeak, Avisoft Bioacoustics, Berlin, Germany) using an external sound card with a sampling rate of 192 kHz (Fire Wire Audio Capture FA-101, Edirol, London, UK) and a portable ultrasonic power amplifier with a frequency range of 1–125 kHz (Avisoft Bioacoustics). The loudspeaker had a frequency range of 1–120 kHz with a relatively flat frequency response (± 12 dB) between 15–80 kHz. It was placed 20 cm away from the end of one arm at a height of 52 cm above the floor. Testing was performed under red light (approximately 11 lux in the center of the maze and between 9 and 12 lux in the arms) in a testing room with no other rats present.

All behavioral tests were conducted between 9–17 h. Prior to each test, behavioral equipment was cleaned using a 0.1 % acetic acid solution followed by drying.

Acoustic Stimuli

The following four acoustic stimuli were used: 50-kHz calls, 50-kHz sine wave tones, 22-kHz calls, and background noise (see Fig. 1). All stimuli were presented for 1 min with a sampling rate of 192 kHz in 16 bit format. Calls and tones were presented at about 69 dB (measured from a distance of 40 cm), and noise was presented with about 50 dB, which corresponds to the background noise during playback of the other stimuli.

50-kHz Calls

Throughout playback, 221 natural 50-kHz calls (total calling time: 15.3 s) were presented. The presentation was composed of a sequence of 3.5 s, which was repeated for 1 min, i.e. 17 times, to assure the presentation of a high number of frequency-modulated calls within a relatively short period of time. Each sequence contained 13 calls (total calling time: 0.90 s). Out of these, 10 were frequency-modulated and 3 were flat, and had the following features: call duration 0.07±0.01 s (mean±SEM); peak frequency: 61.24±1.75 kHz; bandwidth: 4.63±1.21 kHz; frequency modulation: 31.68±4.62 kHz. These calls had been recorded from a male Wistar rat during exploration of a cage containing scents from a cage mate (for setting and recording see: [43]).

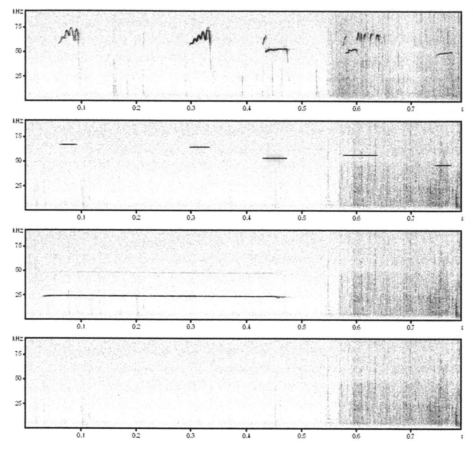

Figure 1. Exemplary spectrograms of the four types of acoustic stimuli presented, namely (from top to down): natural 50-kHz calls, artificial 50-kHz sine wave tones, natural 22-kHz calls, and background noise.

50-kHz Tones

50-kHz sine wave tones were generated with the computer software SASLab Pro (version 4.2, Avisoft Bioacoustics) by replacing all calls through sine wave tones. In detail, each given call was replaced by a sine wave tone with identical duration, frequency, amplitude, etc. Thus, the signal had the same temporal patterning and was identical to the 50-kHz call signal with respect to all call features, apart from the fact that the tones were not amplitude and frequency modulated as the natural 50-kHz calls.

22-kHz Calls

Throughout playback, 29 natural 22-kHz calls (total calling time: 34.25 s) were presented. These calls had the following acoustic parameters: call duration 1.18±0.06 s; peak frequency: 23.61±0.07 kHz; bandwidth: 1.37±0.05 kHz; frequency modulation: 1.90±0.09 kHz. Their presentation was not composed of a repeated sequence, since in case of the long 22-kHz calls potential information, which is contained in temporal patterning is likely lost through sequencing. These calls had been recorded from a male Wistar rat after applications of foot-shocks (for setting and recording see: [10]).

Noise

Since all three acoustic stimuli presented contained background noise, i.e. sounds, which occur when a rat is exploring an arena with bedding, background noise without calls or tones was presented to control for its possible effects.

Experimental Procedure

A given animal was placed onto the central platform of the radial maze, facing the arm opposite to the loudspeaker. After an initial phase of 15 min where no acoustic stimuli were presented (termed habituation), the rat was exposed to three presentations of acoustic stimuli for 1 min, each followed by an inter-stimulus-interval of 10 min.

Between sub-groups of subjects, different orders of stimulation presentation were used to account for the possible impact of sequence effects. In Exp. 1, background noise, 22-kHz calls and 50-kHz calls were used as acoustic stimuli. They were presented in the following orders: a) background noise, b) 22-kHz calls, c) 50-kHz calls (n = 6 rats), or a) background noise, b) 50-kHz calls, c) 22-kHz calls (n = 6). In Exp. 2 and 3, where background noise, 50-kHz sine wave tones and 50-kHz calls were tested used, they were presented either in the order a) background noise, b) 50-kHz sine wave tones, c) 50-kHz calls (Exp. 2: n = 6; Exp. 3: n = 12), or a) background noise, b) 50-kHz calls, c) 50-kHz sine wave tones (Exp. 2: n = 6; Exp. 3: n = 12), or a) 50-kHz calls, b) 50-kHz sine wave tones, c) background noise (Exp. 3: n = 12), or a) 50-kHz calls, background noise, 50-kHz sine wave tones (Exp. 2: n = 7). One animal was excluded from analysis of Exp. 2 due to incorrect presentation of acoustic stimuli.

We abstained from depicting the order of stimulus presentation in detail, since it had no major qualitative effects on the patterns of result, i.e. behavioral responses towards 22-kHz calls and 50-kHz calls were similar over all positions

(Mann-Whitney-U-test for Exp. 1 or Kruskal-Wallis-test for Exp. 2 & 3: all p-values >.100).

Recording and Analysis of Animal Activity

Behavior was monitored by a video camera (Panasonic WV-BP 330/GE, Hamburg, Germany) from about 150 cm above the maze, which fed into DVD recorder (DVR-3100 S, Pioneer, Willich, Germany).

Behavioral analysis was performed in two ways. A trained observer scored the videos for the time spent on the three arms proximal to or distal from the ultrasonic loudspeaker. Furthermore, the total distance travelled (cm), and the number of arm entries into the three proximal or distal arms, were analyzed using an automated video tracking system (Ethovision, Noldus, Wageningen, The Netherlands). For the automated analysis, input filters were activated to avoid an over-estimation of locomotor activity due to head-movements. In more detail, a minimal distance moved of 8 cm was used for the total distance travelled, whereas a minimal distance moved of 3 cm was used for the arm entries.

Recording and Analysis of Ultrasonic Vocalization

Playback of acoustic stimuli and potential ultrasonic calls uttered by the rat under testing were monitored by two UltraSoundGate Condenser Microphones (CM 16; Avisoft Bioacoustics) placed 20 cm away from the maze at a height of 55 cm above the floor. One out of these two was placed next to the loudspeaker, i.e. in front of the three proximal arms, whereas the other one was placed vis-à-vis in front of the three distal arms. These microphones were sensitive to frequencies of 15-180 kHz with a flat frequency response (± 6 dB) between 25–140 kHz, and were connected via an Avisoft UltraSoundGate 416 USB Audio device (Avisoft Bioacoustics) to a personal computer, where acoustic data were displayed in real time by Avisoft RECORDER (version 2.7; Avisoft Bioacoustics), and were recorded with a sampling rate of 214,285 Hz in 16 bit format.

For acoustical analysis, recordings were transferred to SASLab Pro (version 4.38; Avisoft Bioacoustics) and a fast Fourier transform was conducted (512 FFT-length, 100 % frame, Hamming window and 75 % time window overlap). Correspondingly, the spectrograms were produced at 488 Hz of frequency resolution and 0.512 ms of time resolution. The numbers of 22-kHz calls and 50-kHz calls were counted by experienced observers.

Statistical Analysis

Non-parametric statistics were used, since several data sets were not normally distributed as indicated by the Shapiro-Wilk-test. In more detail, the Friedman-test for repeated measurements was calculated to test whether overt or calling behavior is affected by presentation of acoustic stimuli. When appropriate, the Wilcoxon-test was used subsequently to determine whether overt or calling behavior during presentation of a given acoustic stimulus differ in comparison to other acoustic stimuli, or in comparison to phases without presentations of acoustic stimuli. For the last purpose, overt and calling behavior shown in the three min preceding stimulus application was averaged to eliminate habituation effects. Furthermore, the Wilcoxon-test was used to compare the entries into or the time spent on proximal or distal arms of the radial-maze during a given test period. Finally, Spearman correlation coefficients were calculated to test whether individual responses to different acoustic stimuli were stable and whether overt and calling behaviors were related to each other. The exact p-values of 2-tailed testing were taken as measures of effect.

Results

Experiment 1 – Juvenile Rats

This initial experiment was performed to test whether presentation of ultrasonic calls is effective to modify behavior in juvenile rats. Here, we used 22-kHz calls, for which we expected behavioral inhibition, and natural 50-kHz calls, for which we expected activation and orientation towards the source of stimulation.

Locomotor Activity

Locomotor activity of juvenile rats was affected by presentations of acoustic stimuli (see Fig. 2), since the distance travelled was dependent on a) whether acoustic stimuli were presented or not and b) which type of stimulus was presented. In detail, natural 50-kHz calls caused an increase in the distance travelled in comparison to test periods without presentations ($Z = -2.353$, $p = .016$), or to presentation of noise ($Z = -2.934$, $p = .001$). In contrast, locomotor activity was reduced when natural 22-kHz calls were presented, indicated by a decrease when compared versus natural 50-kHz calls ($Z = -2.746$, $p = .003$), and a trend for a decrease in comparison to test periods without presentations ($Z = -1.955$, $p = .055$), but not in comparison to presentation of noise ($Z = -.415$, $p = .734$). Finally, no difference in locomotor activity was found between test periods without presentations and background noise ($Z = -1.070$, $p = .322$).

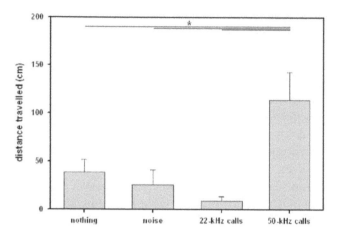

Figure 2. Locomotor activity of juvenile rats in Exp. 1. Bars depict the distance travelled during test phases without acoustic presentation (nothing), presentation of noise (noise), artificial 50-kHz sine wave tones (50-kHz tones), and natural 50-kHz calls (50-kHz calls). Values reflect means±SEM per minute. Animals of all stimulus orders were collapsed, i.e. n = 12. Comparisons with p<.05 are marked with asterisks: *.

Stimulus-Directed Locomotor Activity

As expected, only natural 50-kHz calls, but not natural 22-kHz calls, induced approach behavior. Thus, during presentations of 50-kHz calls animals entered the three proximal arms in front of the loudspeaker more often than the three distal ones (Z = −2.456, p = .016) and spent more time in the former (Z = −3.059, p<.001). No preference was observed during playback of noise or natural 22-kHz calls (all p-values >.100). Remarkably, approach behavior during playback of 50-kHz calls was evident despite the fact that the animals showed an a-priori preference for the distal arms, indicated by more entries into distal arms than in proximal ones and the fact that animals spent more time in the distal arms than proximal ones during habituation (Z = −2.185, p = .026 and Z = −2.510, p = .009, respectively) and after cessation of noise (Z = −1.720, p = .084 and Z = −2.134, p = .032, respectively). After playback of 22-kHz calls, no preference was found (all p-values >.100), whereas animals tended to stay longer in proximal arms than in distal ones after presentation of 50-kHz calls (Z = −1.805, p = .076; arm entries: Z = −1.660, p = .110). When comparing the time spent on proximal arms during playback of 22-kHz calls and 50-kHz calls, it was found that animals spent more time on proximal arms during playback of 50-kHz calls (Z = −2.589, p = .007; see Fig. 3). This stimulus-dependent difference was also evident after cessation of acoustic stimuli (Z = −2.040, p = .042), indicating that 50-kHz calls can induce a sustained preference for the source of playback.

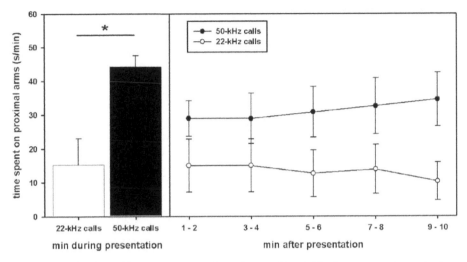

Figure 3. Stimulus-directed locomotor activity of juvenile rats in Exp. 1. The time spent on the proximal arms in front of the loudspeaker is given for playback of natural 22-kHz calls (white bar) and natural 50-kHz calls (black bar) is depicted on the left. On the right, the time spent on the proximal arms in front of the loudspeaker is given for the 10 min after cessation of playback of natural 22-kHz calls (open symbols) and natural 50-kHz calls (filled symbols). Values reflect means±SEM per minute. In both cases, animals of all stimulus orders were collapsed, i.e. n = 12. Comparisons with p<.05 are marked with asterisks: *.

Ultrasonic Calling

During testing, 7 out of 12 animals emitted some 50-kHz calls (1.75±0.65, i.e. 0.02±0.01 per min). However, none of them emitted 50-kHz calls during presentation of 50-kHz calls, or 22-kHz calls, and only one animal emitted a single call during presentation of noise, meaning that calls were predominantly emitted during inter-stimulus-intervals (not shown in detail).

22-kHz calls were not observed. However, calls with a similar shape and a long duration up to 900 ms, but an atypical high frequency, were found in one animal, which emitted 15 calls after cessation of presentations of 50-kHz calls (not shown in detail). Remarkably, it emitted also 50-kHz calls.

Experiment 2 – Juvenile Rats

Here, we again used juvenile subjects and tested whether behavioral activation and approach might not only be elicited by natural 50-kHz calls, but also by artificial 50-kHz sine wave tones which had the same temporal patterning and were identical to 50-kHz calls with respect to all call features, apart from the fact that the tones were not amplitude and frequency modulated.

Locomotor Activity

In replication of Exp. 1, it was found that 50-kHz calls caused an increase in the distance travelled in comparison to test periods without presentations (Z = –3.662, p<.001), or to presentation of noise (Z = –3.662, p<.001; see Fig. 4). In contrast, playback of 50-kHz tones did not induce locomotor activation, and locomotor activity during presentation of 50-kHz tones was lower as during presentation of 50-kHz calls (Z = –3.340, p<.001; all other p-values >.100). Finally, no difference in locomotor activity was found between test periods without presentations and background noise (Z = –1.046, p = .312).

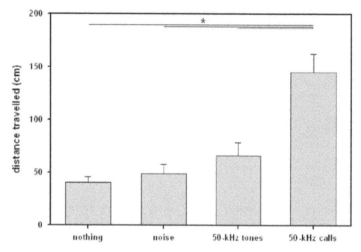

Figure 4. Locomotor activity of juvenile rats in Exp. 2. Bars depict the distance travelled during test phases without acoustic presentation (nothing), presentation of noise (noise), artificial 50-kHz sine wave tones (50-kHz tones), and natural 50-kHz calls (50-kHz calls). Values reflect means±SEM per minute. Animals of all stimulus orders were collapsed, i.e. n = 19. Comparisons with p<.05 are marked with asterisks: *.

Stimulus-Directed Locomotor Activity

Furthermore, it was found that locomotor activity was stimulus-directed during both, presentation of 50-kHz tones and natural 50-kHz calls (see Fig. 5), since the animals entered the three proximal arms in front of the loudspeaker more often than the distal ones (50-kHz tones: Z = –2.012, p = .055; 50-kHz calls: Z = –3.572, p<.001). Furthermore, they spent more time on the proximal arms than on the distal ones (50-kHz tones: Z = –3.575, p<.001; 50-kHz calls: Z = –3.823, p<.001). Such preferences were not observed during test periods without presentations, or during presentation of noise, except for a trend for a longer time spent on proximal arms relatively to distal ones after the cessation of presentation of 50-kHz calls (Z = –1.811, p = .073; all other p-values >.100).

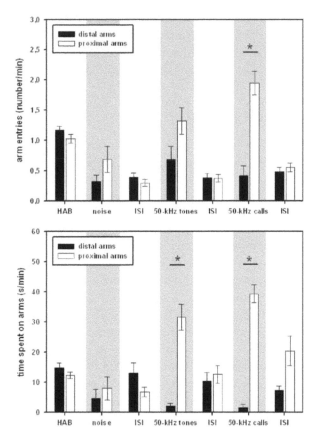

Figure 5. Stimulus-directed locomotor activity of juvenile rats in Exp. 2. The number of entries into the distal (black bars) or proximal (white bars) arms from the loudspeaker is given for habituation (HAB), inter-stimulus-intervals (ISI), and playback of acoustic stimuli, i.e. natural 50-kHz calls (50-kHz calls), artificial 50-kHz sine wave tones (50-kHz tones), and background noise (noise) in the upper figure. The time spent on the distal (black bars) or proximal (white bars) arms from the loudspeaker is given for habituation (HAB), inter-stimulus-intervals (ISI), and playback of acoustic stimuli, i.e. natural 50-kHz calls (50-kHz calls), artificial 50-kHz sine wave tones (50-kHz tones), and background noise (noise) in the bottom figure. Values reflect means±SEM per minute. In both cases, animals of all stimulus orders were collapsed, i.e. n = 19. Comparisons with p<.05 are marked with asterisks: *.

Ultrasonic Calling

During testing, 10 out of 19 animals emitted 50-kHz calls. However, call rates were very low (1.42±0.58, i.e. 0.03±0.01 per min), and none of them emitted 50-kHz calls during presentation of 50-kHz tones or 50-kHz calls. Solely 1 animal emitted 1 single call during presentation of noise, meaning that 50-kHz calls were predominantly emitted during ISIs (not shown in detail).

22-kHz calls were not observed. However, calls with a similar shape and a long duration up to 900 ms, but an atypical high frequency, were found in some few animals. Throughout the whole testing period, 3 out of 19 animals emitted them (1, 4 and 22 calls). Calls were primarily emitted during the presentations of 50-kHz tones or 50-kHz calls and after cessation of presentations (not shown in detail). Remarkably, 2 out of the 3 animals also emitted 50-kHz calls.

Experiment 3 – Adult Animals

In this final experiment, we used the same approach as in Exp.2, and asked whether 50-kHz calls or 50-kHz sine wave tones might also be effective when presented to adult rats.

Locomotor Activity

As in juvenile rats, locomotor activity was dependent on on a) whether acoustic stimuli were presented or not and b) which type of stimulus was presented (see Fig. 6). In detail, 50-kHz calls caused an increase in the distance travelled in comparison to test periods without presentations ($Z = -.3833$, p<.001), or to noise ($Z = -3.976$, p<.001). Furthermore, a similar increase in the distance travelled was observed when 50-kHz tones were presented (in comparison to periods without presentations: $Z = -3.620$, p<.001; in comparison to presentation of noise: $Z = -3.548$, p<.001). Remarkably, the distance travelled did not differ between presentations of 50-kHz tones and 50-kHz calls ($Z = -.131$, p = .903). Finally, no difference in locomotor activity was found between test periods without presentations and background noise ($Z = -1.456$, p = .150).

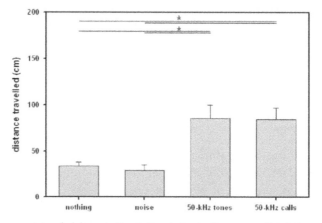

Figure 6. Locomotor activity of adult rats in Exp. 3. Bars depict the distance travelled during test phases without acoustic presentation (nothing), presentation of noise (noise), artificial 50-kHz sine wave tones (50-kHz tones), and natural 50-kHz calls (50-kHz calls). Values reflect means±SEM per minute. Animals of all stimulus orders were collapsed, i.e. n = 36. Comparisons with p<.05 are marked with asterisks: *.

Stimulus-Directed Locomotor Activity

Locomotor activity was stimulus-directed during presentations of 50-kHz tones and 50-kHz calls (see Fig. 7), since the animals entered the three proximal arms in front of the loudspeaker more often than the three distal ones ($Z = -4.110$, $p = .001$ and $Z = -3.155$, $p<.001$, respectively). Also, they spent more time on the proximal arms (50-kHz tones: $Z = -2.575$, $p = .008$; 50-kHz calls: $Z = -2.516$, $p = .010$). Such preferences were not observed during test periods without presentations or presentation of noise (all p-values >.100).

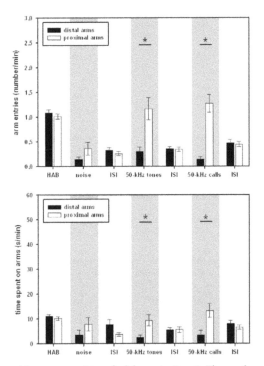

Figure 7. Stimulus-directed locomotor activity of adult rats in Exp. 3. The number of entries into the distal (black bars) or proximal (white bars) arms from the loudspeaker is given for habituation (HAB), inter-stimulus-intervals (ISI), and playback of acoustic stimuli, i.e. natural 50-kHz calls (50-kHz calls), artificial 50-kHz sine wave tones (50-kHz tones), and background noise (noise) in the upper figure. The time spent on the distal (black bars) or proximal (white bars) arms from the loudspeaker is given for habituation (HAB), inter-stimulus-intervals (ISI), and playback of acoustic stimuli, i.e. natural 50-kHz calls (50-kHz calls), artificial 50-kHz sine wave tones (50-kHz tones), and background noise (noise) in the bottom figure. Values reflect means±SEM per minute. In both cases, animals of all stimulus orders were collapsed, i.e. n = 36. Comparisons with p<.05 are marked with asterisks: *.

Ultrasonic Calling

During testing, 26 out of 36 animals emitted 50-kHz calls (5.44±2.49, i.e. 0.11±0.05 per min). Out of these, 8 animals emitted 50-kHz calls during presentation of 50-kHz tones or 50-kHz calls, but none animal emitted 50-kHz calls during presentation of noise. Remarkably, 50-kHz calling was affected by presentations of acoustic stimuli (see Fig. 8). Call emission was higher during presentations of 50-kHz calls than during testing periods without presentation ($Z = -2.157$, $p = .047$) or presentation of noise ($Z = -2.410$, $p = .016$), whereas call emission during presentations of 50-kHz tones did not differ from any other test period (all p-values $>.100$), indicating that only playback of 50-kHz calls induced 50-kHz calling. Finally, no difference in calling behavior was found between test periods without presentations and background noise ($Z = -1.414$, $p = .500$).

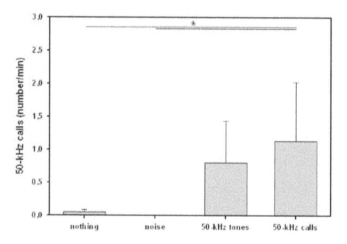

Figure 8. Ultrasonic calling of adult rats in Exp. 3. Bars depict the number of 50-kHz calls emitted by the subject under study during test phases without acoustic presentation (nothing), presentation of noise (noise), artificial 50-kHz sine wave tones (50-kHz tones), and natural 50-kHz calls (50-kHz calls). Values reflect means±SEM per minute. Animals of all stimulus orders were collapsed, i.e. n = 36. Comparisons with p<.05 are marked with asterisks: *

Interestingly, 50-kHz calling was related to activity and approach behavior during presentations of 50-kHz tones and 50-kHz calls. In detail, during presentation of 50-kHz tones the number of 50-kHz calls emitted was positively correlated with the distance travelled (rho = .394, p = .017), the number of entries in proximal arms (rho = .404, p = .014) and the time spent there (rho = .346, p = .039), but not with the number of entries in distal arms (rho = .043, p = .803) and the time spent there (rho = .314, p = .062). During presentations of 50-kHz calls, the number of 50-kHz calls emitted by the subject under study was positively correlated with the distance travelled (rho = .345, p = .039), the number of entries in proximal arms (rho = .386, p = .020) and tended to correlate with the time spent there (rho = .299, p = .076), but no with the number of entries in distal arms (rho = .017, p = .922) and the time spent there (rho = -.147, p = .392) were observed.

No correlations between 50-kHz calling and locomotor activity and the direction of locomotor activity were found during habituation (all p-values >.050).

22-kHz calls were very rarely observed. Throughout the whole testing period, only 2 out of 36 animals emitted them. One of them emitted 9 calls after cessation of the presentation of 50-kHz tones, the other one emitted 2 calls after cessation of the presentation of 50-kHz calls (not shown in detail). Remarkably, both animals emitted not only 22-kHz calls, but also 50-kHz calls. Actually, the first one displayed the highest number of 50-kHz calls throughout the whole testing period (90 calls), but also throughout the presentations 50-kHz tones (22 calls) and 50-kHz calls (32 calls).

Discussion

Our results demonstrate for the first time that 50-kHz calls can induce approach behavior and ultrasonic calling in non-sexual contexts, whereas 22-kHz calls induced a reduction in locomotor activity.

Playback of 22-kHz Calls Induce Behavioral Inhibition

The present findings are in line with several previous ones, which have already shown that 22-kHz calls can activate the fight/flight/freeze system. Dependent on the strain of the receiver, 22-kHz calls can induce behavioral inhibition [48]–[51], or bursts of locomotor running and jumping, which are characteristic of defence behavior [49], [50], [52], [53]. However, it has to be noted that studies using natural 22-kHz calls obtained only a moderate reduction of locomotor activity [48], [51], [59], which is in line with the relatively weak effects of 22-kHz calls found here. From these results, one should not conclude that 22-kHz calls do not provide important signals for the recipient; rather, one should assume that their salience depends on additional features like a given social context [5], or whether they are linked to critical experiences [59].

Playback of 50-kHz Calls Can Induce Activation and Approach

Studies on the behavioral effects of 50-kHz calling using playback methods were predominantly conducted in the sexual context. Here, it was found that darting behavior and approaches toward the partner increased in frequency when the female was devocalized, but decreased when tape recorded female ultrasonic calls were presented [56], [57]. With respect to male USV, it was shown that devocalization of male rats resulted in a reduction of female proceptive behavior [61], and playback of 50-kHz calls restored proceptive behavior in oestrus females [23], [25], [27].

The few studies, which were conducted in a non-sexual context, however, obtained very weak or no playback-induced effects on overt behavior. Thus, early studies using artificial ultrasonic stimuli observed Preyer's reflex [54], or a suppression of instrumental bar pressing and bradycardia [55], possibly reflecting an unspecific orienting response. Finally, a recent study by Endres et al. [59], did not observe any change in overt behavioral activity when natural 50-kHz calls were presented in comparison to other acoustic stimuli, like white noise or even 22-kHz calls. Therefore, the present study is the first one, which clearly shows that 50-kHz calls can affect overt and calling behavior in a non-sexual context. In accordance to the hypothesis that 50-kHz calls serve communicative purposes [44], [62], [63], we found that animals increase locomotor activity and approach the source of the stimulus, resembling mothers when searching for their pups in response to isolation-induced pup calls [45], [46], [47].

Furthermore, we showed that playback of 50-kHz calls can elicit ultrasonic calling by the recipient subject, which is in line with findings by White et al. [58] showing that male 50-kHz calls can elevate female calling. Thus, the present findings clearly indicate that the communicative value of 50-kHz calls is not restricted to sexual interactions. Therefore, it can be concluded that differences between sexual and non-sexual contexts are not responsible for the conflicting findings. Possible reasons for the lack of evidence in previous studies might be due to the type of stimulus material and playback technology used in the early playback work [54], [55], or the experimental setting used in the study of Endres et al. [59], who mounted their loudspeaker above the testing arena and not at the side, as done here. Possibly, 50-kHz signals coming from the horizontal axis might provide a more naturalistic signal for the recipient than calls coming from above.

Frequency Modulation is not Necessary for Eliciting Approach Behavior

The fact that 50-kHz calls induced approach behavior clearly indicates that these calls were appetitive, which is in line with findings by Burgdorf et al. [32] who showed that rats show instrumental behavior to receive 50-kHz calls. There, frequency-modulated, but not flat 50-kHz calls were effective, whereas the present results demonstrate that 50-kHz signals with and without amplitude and frequency modulation are appetitive, since artificial 50-kHz sine wave tones also induced approach behavior. Despite the fact that natural 50-kHz calls tended to be more efficient in eliciting behavioral changes, amplitude and frequency modulation is apparently not a necessary prerequisite for the appetitive value of 50-kHz calls. Therefore, the present results are more in accordance with the assumption that a whole bundle of call features is responsible for the information, which is conveyed by such calls. We suggest, therefore, a compensatory model for 50-kHz calls, which states that the whole signal information is not lost when a specific call feature is missing, what would be predicted on the basis of the alternative conjunctive model.

Alternatively, one could assume that both, flat and frequency modulated calls, might be appetitive, but that the value of the latter is perhaps higher than that one of flat calls, a difference which is more likely to be detected in tests, like the one used by Burgdorf et al. [32], where the animal can actively chose between playback of different call varieties. Another explanation is that peak frequency rather than frequency-modulation is critical for the appetitive value of 50-kHz calls, since Burgdorf et al. [32] showed that frequency-modulated and flat calls also differ in their peak frequency. In the present study, only the amplitude and frequency modulation of calls was removed, but mean peak frequency remained unchanged, meaning that the 50-kHz sine wave tones used here had a peak frequency, which is typical for frequency-modulated calls. Actually, Brudzynksi [64] has suggested that, apart from call number, peak frequency is involved in coding the quantitative aspect of the sign function of 50-kHz calls, since peak frequency can be modulated by pharmacological agents, like glutamate [65].

Juvenile Rats Respond More Strongly to 50-kHz Calls than Adult Rats

Furthermore, we found that effects on overt behavior were more pronounced in juvenile rats than in adult rats. This age-related difference is even more impressive, when considering the relatively small number of young animals and the fact that the effect was evident irrespective of whether 22-kHz calls were presented in the same test or not. The difference in approach behavior between juvenile and adult rats is possibly reflecting a decrease in social interest in function of ageing. In fact, a reduced level of gregariousness among older individuals was consistently found in mammals. For instance, in a wide variety of primate species, aging leads to active withdrawal from social interactions and an increase in time spent alone [66]–[68]. Similar changes in function of age were also found in rats and mice. Thus, Salchner et al. [69] were able to show that aged rats spent considerably less time in active social interaction than young rats. Recently, Moles et al. [70] replicated this finding in mice. Interestingly, they did not only observe a decrease in the time spent investigating the partner, but also in the number of USV.

Furthermore, the stronger overt behavioral response in juvenile rats is in accordance with observations that 50-kHz calls occur predominantly in juvenile rats [37]. However, it remains unclear why young animals do not vocalize at all during playback of 50-kHz calls, whereas adult rats displayed ultrasonic calling in response to playback. One point, which might be of relevance in this context, is that the 50-kHz calls presented where emitted by adult rats, and it seems to be possible that call characteristics may convey information about age and status. Apart from these differences between juvenile and adult rats, it was observed that adult rats responded similarly to 50-kHz sine wave tones as to natural 50-kHz calls, whereas the response toward the artificial tones was not as strong as toward the natural calls in young animals. This difference might be due to a reduced acoustic sensitivity and plasticity in adult animals [71].

50-kHz Ultrasonic Calling and Social Approach

Rats are gregarious. For instance, two rats placed together in a large chamber spend substantially more time together than would be expected by chance, and are more attracted to other rats than to physical objects [72], [73]. Obviously, social approach is crucial for establishing and maintaining relationships among individuals. The present findings indicate that the emission of 50-kHz calls is an important element in the evolvement of social relationships in rats. In fact, 50-kHz calls are typically emitted during social interactions, like reproductive behavior [21], [23], [25]–[28], juvenile play [19], [20] and tickling [36]–[40]. That emission of 50-kHz calls is functional for these behaviors is indicated by studies showing that deafening or devocalizing rats can affect reproductive behavior [23], [25], [27], [28], [56], [61] and reduces rough-and-tumble play [74]. Correspondingly, it was found that animals prefer to spend more time with other animals that vocalize a lot rather than with those that do not [75]. Furthermore, rats emit 50-kHz calls when entering areas where social contact has previously occurred [22], [24], [44], [76], [77]. Remarkably, the present findings nicely fit into earlier studies where it was shown that adult rats emit 50-kHz calls after separation from the cage mate, indicating that such calling serves to (re)establish or keep contact [43]. Similar conclusions can be drawn for mice, where USV was found during mating and social exploration [70], [78]–[81]. Interestingly, Panksepp et al. [80] observed that high-frequency calling in mice is positively correlated with social investigation. Furthermore, Moles and D`Amato [79] have shown that social investigation and the number of ultrasonic calls can be modulated by manipulating the attractiveness of the test partner. They have suggested, therefore, that ultrasonic calls facilitate proximity between animals, which helps to acquire relevant social information.

The study of social approach in laboratory animals can help to reveal biochemical, genetic and environmental factors underlying neuropsychiatric disorders such as depression, autism and Rett syndrome, since these are characterized, among others, by social deficits and loss of desire to engage in social interactions [82]. Bearing in mind the wealth of evidence implicating 50-kHz calls as a key element of social interactions in rats, it is noteworthy that the measurement of behavioral responses toward playback of 50-kHz calls provides a rather unique opportunity to study the determinants of social interest by using a standardized non-social test, i.e. without confounding effects of a partner. For instance, it is possible to model two core symptoms of the autistic syndrome, namely lack of social interest and communicative deficits [83], [84].

Conclusion

The present findings clearly show that 50-kHz calls can induce approach behavior and ultrasonic calling in male rats. Thus, the hypothesis that such 50-kHz calls serve for communicative purposes, for example, to (re)establish or to keep contact with conspecifics, is supported.

Acknowledgements

The authors wish to thank Benedikt T. Bedenk, Moriah Hülse-Matia and Claudia Lucas for their support in data acquisition.

Authors' Contributions

Conceived and designed the experiments: MW RS. Performed the experiments: MW. Analyzed the data: MW. Contributed reagents/materials/analysis tools: RS. Wrote the paper: MW RS.

References

1. Constantini F, D'Amato F (2006) Ultrasonic vocalizations in mice and rats: social contexts and functions. Acta Zool Sinica 52: 619–633.

2. Knutson B, Burgdorf J, Panksepp J (2002) Ultrasonic vocalizations as indices of affective states in rats. Psychol Bull 28: 961–977.

3. Portfors CV (2007) Types and functions of ultrasonic vocalizations in laboratory rats and mice. J Am Assoc Lab Anim Sci 46: 28–34.

4. Hofer MA (1996) Multiple regulators of ultrasonic vocalization in the infant rat. Psychoneuroendocrinology 21: 203–217.

5. Blanchard RJ, Blanchard DC, Agullana R, Weiss SM (1991) Twenty-two kHz alarm cries to presentation of a predator, by laboratory rats living in visible burrow systems. Physiol Behav 50: 967–72.

6. Borta A, Wöhr M, Schwarting RKW (2006) Rat ultrasonic vocalization in aversively motivated situations and the role of individual differences in anxiety-related behavior. Behav Brain Res 166: 271–280.

7. Cuomo V, de Salvia MA, Maselli MA, Renna G, Racagni G (1998) Ultrasonic vocalization in response to unavoidable aversive stimuli in rats: effects of benzodiazepines. Life Sci 43: 485–491.

8. Jelen P, Soltysik S, Zagrodzka J (2003) 22-kHz ultrasonic vocalization in rats as an index of anxiety but not fear: behavioral and pharmacological modulation of affective state. Behav Brain Res 141: 63–72.

9. Tonue T, Ashida Y, Makino H, Hata H (1986) Inhibition of shock-induced ultrasonic vocalization by opioid peptides in the rat: a psychotropic effect. Psychoneuroendocrinology 11: 177–184.

10. Wöhr M, Borta A, Schwarting RKW (2005) Overt behavior and ultrasonic vocalization in a fear conditioning paradigm-a dose-response study in the rat. Neurobiol Learn Mem 84: 228–240.

11. Kaltwasser MT (1990) Acoustic signaling in the black rat (Rattus rattus). J Comp Psychol 104: 227–232.

12. Sales GD (1972a) Ultrasound and aggressive behavior in rats and other small mammals. Anim Behav 20: 88–100.

13. Barros HMT, Miczek KA (1996) Withdrawal from oral cocaine in rats: ultrasonic vocalizations and tactile startle. Psychopharmacology (Berl) 125: 379–384.

14. Covington HE, Miczek KA (2003) Vocalizations during withdrawal from opiates and cocaine: possible expressions of affective distress. Eur J Pharmacol 467: 1–13.

15. Brudzynski SM, Ociepa D (1992) Ultrasonic vocalization of laboratory rats in response to handling and touch. Physiol Behav 52: 655–660.

16. Francis RL (1977) 22-kHz calls by isolated rats. Nature 265: 236–238.

17. Sanchez C (2003) Stress-induced vocalisation in adult animals. A valid model of anxiety? Eur J Pharmacol 463: 133–143.

18. Miczek KA, Weerts EM, Vivian JA, Barros HM (1995) Aggression, anxiety and vocalizations in animals: GABAA and 5-HT anxiolytic. Psychopharmacology (Berl) 121: 38–56.

19. Brunelli SA, Nie R, Whipple C, Winiger V, Hofer MA, et al. (2006) The effects of selective breeding for infant ultrasonic vocalizations on play behavior in juvenile rats. Physiol Behav 87: 527–536.

20. Knutson B, Burgdorf J, Panksepp J (1998) Anticipation of play elicits high-frequency ultrasonic vocalizations in young rats. J Comp Psychol 112: 65–73.

21. Barfield RJ, Auerbach P, Geyer LA, McIntosh TK (1979) Ultrasonic vocalizations in rat sexual behavior. Am Zool 19: 469–480.

22. Bialy M, Rydz M, Kaczmarek L (2000) Precontact 50-kHz vocalizations in male rats during acquisition of sexual experience. Behav Neurosci 114: 983–90.

23. Geyer LA, Barfield RJ (1978) Influence of gonadal hormones and sexual behavior on ultrasonic vocalization in rats. I. Treatment of females. J Comp Physiol Psychol 92: 436–446.

24. McGinnis MY, Vakulenko M (2003) Characterization of 50-kHz ultrasonic vocalizations in male and female rats. Physiol Behav 80: 81–88.

25. McIntosh TK, Barfield RJ, Geyer LA (1978) Ultrasonic vocalisations facilitate sexual behavior of female rats. Nature 272: 163–164.

26. Sales GD (1972b) Ultrasound and mating behavior in rodents with some observations on other behavioral situations. J Zool 168: 149–164.

27. White NR, Barfield RJ (1990) Effects of male pre-ejaculatory vocalizations on female receptive behavior in the rat (Rattus norvegicus). J Comp Psychol 104: 140–146.

28. White NR, Cagiano R, Moises AU, Barfield RJ (1990) Changes in mating vocalizations over the ejaculatory series in rats (Rattus norvegicus). J Comp Psychol 104: 255–62.

29. Burgdorf J, Knutson B, Panksepp J (2000) Anticipation of rewarding electrical brain stimulation evokes ultrasonic vocalization in rats. Behav Neurosci 114: 320–327.

30. Burgdorf J, Wood PL, Kroes RA, Moskal JR, Panksepp J (2007) Neurobiology of 50-kHz ultrasonic vocalizations in rats: electrode mapping, lesion and pharmacological studies. Behav Brain Res 182: 274–283.

31. Burgdorf J, Knutson B, Panksepp J, Ikemoto S (2001) Nucleus accumbens amphetamine microinjections unconditionally elicit 50-kHz ultrasonic vocalizations in rats. Behav Neurosci 115: 940–4.

32. Burgdorf J, Kroes RA, Moskal JR, Pfaus JG, Brudzynski SM, et al. Ultrasonic vocalizations of rats during mating, play, and aggression: behavioral concomitants, relationship to reward, and self-administration of playback. J Comp Psychol. In press.

33. Knutson B, Burgdorf J, Panksepp J (1999) High-frequency ultrasonic vocalizations index conditioned pharmacological reward in rats. Physiol Behav 66: 639–643.

34. Thompson B, Leonard KC, Brudzynski SM (2006) Amphetamine-induced 50 kHz calls from rat nucleus accumbens: A quantitative mapping study and acoustic analysis. Behav Brain Res 168: 64–73.

35. Wintink AJ, Brudzynski SM (2001) The related roles of dopamine and glutamate in the initiation of 50-kHz ultrasonic calls in adult rats. Pharmacol Biochem Behav 70: 317–323.

36. Burgdorf J, Panksepp J (2001) Tickling induces reward in adolescent rats. Physiol Behav 72: 167–173.

37. Panksepp J, Burgdorf J (1999) Laughing rats? Playful tickling arouses high frequency ultrasonic chirping in young rodents. In: Hameroff S, Chalmers C, Kazniak A, editors. Toward a science of consciousness III. Camebridge, Massachusetts: MIT Press. pp. 231–244.

38. Panksepp J, Burgdorf J (2000) 50-kHz chirping (laughter?) in response to conditioned and unconditioned tickle-induced reward in rats: effects of social housing and genetic variables. Behav Brain Res 115: 25–38.

39. Panksepp J, Burgdorf J (2003) "Laughing" rats and the evolutionary antecedents of human joy? Physiol Behav 79: 533–547.

40. Schwarting RKW, Jegan N, Wöhr M (2007) Situational factors, conditions and individual variables which can determine ultrasonic vocalizations in male adult Wistar rats. Behav Brain Res 182: 208–222.

41. Burgdorf J, Knuston B, Panksepp J, Shippenberg TS (2001) Evaluation of rat ultrasonic vocalizations as predictors of the conditioned aversive effects of drugs. Psychopharmacology (Berl) 155: 35–42.

42. Burgdorf J, Panksepp J (2006) The neurobiology of positive emotions. Neurosci Biobehav Rev 30: 173–187.

43. Wöhr M, Houx B, Schwarting RKW, Spruijt BEffects of experience and context on 50-kHz vocalizations in rats. Physiol Behav. In press. doi: 10.1016/j.physbeh.2007.11.031.

44. Brudzynski SM, Pniak A (2002) Social contacts and production of 50-kHz short ultrasonic calls in adult rats. J Comp Psychol 116: 73–82.

45. Allin JT, Banks EM (1972) Functional aspects of ultrasound production by infant albino rats (Rattus norvegicus). Anim Behav 20: 175–185.

46. Smotherman WP, Bell RW, Starzec J, Elias J, Zachman TA (1974) Maternal responses to infant vocalizations and olfactory cues in rats and mice. Behav Biol 12: 55–66.

47. Wöhr M, Schwarting RKWMaternal care, isolation-induced infant ultrasonic calling, and their relations to adult anxiety-related behavior in the rat. Behav Neurosci. In press.

48. Brudzynski SM, Chiu EMC (1995) Behavioral responses of laboratory rats to playback of 22 kHz ultrasonic calls. Physiol Behav 57: 1039–1044.

49. Commisssaris RL, Palmer A, Neophytou SI, Graham M, Beckett SRG, et al. (2000) Acoustically elicited behaviors in Lister hooded and Wistar rats. Physiol Behav 68: 521–531.

50. Neophytou SI, Graham M, Williams J, Aspley S, Marsden CA, et al. (2000) Strain differences to the effects of aversive frequency ultrasound on behavior and brain topography of c-fos expression in the rat. Brain Res 854: 158–164.

51. Sales GD (1991) The effect of 22 kHz calls and artificial 38 kHz signals on activity in rats. Behav Processes 24: 83–93.

52. Beckett SRG, Aspley S, Graham M, Marsden CA (1996) Pharmacological manipulation of ultrasound induced defence behavior in the rat. Psychopharmacology (Berl) 127: 384–390.

53. Beckett SRG, Duxon MS, Aspley S, Marsden CA (1997) Central c-fos expression following 20kHz/ultrasound induced defence behavior in the rat. Brain Res Bull 42: 421–426.

54. Schleidt (1952) Reaktionen von Tönen hoher Frequenz bei Nagern. Naturwissenschaften 39: 69–70.

55. Thomas DA, Haroutunian V, Barfield RJ (1981) Behavioral and physiological habituation to an ultrasonic stimulus. Bull Psychonomic Soc 17: 279–282.

56. White NR, Barfield RJ (1987) Role of the ultrasonic vocalization of the female rat (Rattus norvegicus) in sexual behavior. J Comp Psychol 101: 73–81.

57. White NR, Barfield RJ (1989) Playback of female rat ultrasonic vocalizations during sexual behavior. Physiol Behav 45: 229–233.

58. White NR, Gonzales RN, Barfield RJ (1993) Do vocalizations of the male rat elicit calling from the female? Behav Neural Biol 59: 16–78.

59. Endres T, Widmann K, Fendt M (2007) Are rats predisposed to learn 22 kHz calls as danger-predicting signals? Behav Brain Res 185: 69–75.

60. Görisch J, Schwarting RKW (2006) Wistar rats with high versus low rearing activity differ in radial maze performance. Neurobiol Learn Mem 86: 175–187.

61. Thomas DA, Howard SB, Barfield RJ (1982) Male-produced postejaculatory vocalizations and the mating behavior of estrous female rats. Behav Neural Biol 36: 403–410.

62. Sales GD, Pye D (1974) Ultrasonic communication by animals. New York: Wiley.

63. Smith JW (1979) The study of ultrasonic communication. Am Zool 18: 531–538.

64. Brudzynski SM (2005) Principles of rat communication: quantitative parameters of ultrasonic calls in rats. Behav Genet 35: 85–92.

65. Fu XW, Brudzynski SM (1994) High-frequency ultrasonic vocalization induced by intracerebral glutamate in rats. Pharmacol Biochem Behav 49: 835–841.

66. Hauser MD, Tyrell G (1984) Old age and its behavioral manifestations: a study on two species of macaque. Folia Primatol 43: 24–35.

67. Picq JL (1992) Aging and social behavior in captivity in Microcebus murinus. Folia Primatol 59: 217–220.

68. Veenema HC, van Hooff JA, Gispen WH, Spruijt BM (2001) Increase rigidity with age in social behavior of Java-monkeys (macaca fascicularis). Neurobiol Aging 22: 273–281.

69. Salchner P, Lubec G, Singewald N (2004) Decreased social interaction in aged rats may not reflect changes in anxiety-related behavior. Behav Brain Res 51: 1–8.

70. Moles A, Constantini F, Garbugino L, Zanettini C, D`Amato FR (2007) Ultrasonic vocalization emitted during dyadic interactions in female mice: a possible index of sociability? Behav Brain Res 182: 223–230.

71. Keuroghlian AS, Knudsen EI (2007) Adaptive auditory plasticity in developing and adult animals. Prog Neurobiol 82: 109–121.

72. Latané B (1969) Gregariousness and fear in labpratory rats. J Exp Soc Psychol 5: 61–69.

73. Latané B, Glass D (1968) Social and nonsocial attraction in rats. J Pers Soc Psychol 9: 142–146.

74. Siviy SM, Panksepp J (1987) Sensory modulation of juvenile play in rats. Dev Psychobiol 20: 39–55.

75. Panksepp J, Gordon N, Burgdorf J (2002) Empathy and the action-perception resonances of basic socio-emotional systems of the brain. Behav Brain Sci 25: 43.

76. Tornatzky W, Miczek (1994) Behavioral and autonomic responses to intermittent social stress: differential protection by clonidine and metoprolol. Psychopharmacology (Berl) 116: 346–356.

77. Tornatzky W, Miczek (1995) Alcohol, anxiolytics and social stress in rats. Psychopharmacology (Berl) 121: 135–144.

78. Holy TE, Guo Z (2005) Ultrasonic Songs of Male Mice. PLoS Biol 3: e386 doi:10.1371/journal.pbio.0030386.

79. Moles A, D`Amato FR (2000) Ultrasonic vocalization by female mice in the presence of a conspecific carrying food cues. Anim Behav 60: 689–694.

80. Panksepp JB, Jochman KA, Kim JU, Koy JJ, Wilson ED, et al. (2007) Affiliative behavior, ultrasonic communication and social reward are influenced by genetic variation in adolescent mice. PLOS One 2: e351. doi: 10.1371/journal.pone.0000351.

81. White NR, Prasad N, Barfield RF, Nyby JG (1998) 40- and 70-kHz vocalizations of mice (Mus musculus) during copulation. Physiol Behav 63: 467–473.

82. Ricceri L, Moles A, Crawley J (2007) Behavioral phenotyping of mouse models of neurodevelopmental disorders: relevant social behavior patterns across the life span. Behav Brain Res 176: 40–52.

83. Crawley JN (2004) Designing mous behavioral tasks relevant to autistic-like behaviors. Ment Retard Dev Disabil Res Rev 10: 248–258.

84. Moy SS, Nadler JJ, Magnuson TR, Crawley JN (2006) Mouse models of autism spectrum disorders: the challenge for behavioral genetics. Am J Med Genet C Semin Med Genet 142C: 40–51.

CITATION

Wöhr M and Schwarting RKW. Ultrasonic Communication in Rats: Can Playback of 50-kHz Calls Induce Approach Behavior? PLoS ONE 2(12): e1365. doi:10.1371/journal.pone.0001365. Copyright © 2007 Wöhr, Schwarting. Originally published under the Creative Commons Attribution License, http://creativecommons.org/licenses/by/3.0/

Supplementary Feeding Affects the Breeding Behavior of Male European Treefrogs (*Hyla arborea*)

Ivonne Meuche and T. Ulmar Grafe

ABSTRACT

Background

We investigated the effects of energetic constraints on the breeding behavior of male European treefrogs Hyla arborea and how calling males allocated additional energy supplied by feeding experiments.

Results

Presence in the chorus was energetically costly indicated by both fed and unfed males losing weight. Males that were supplied with additional energy did not show longer chorus tenure. Instead, fed males returned sooner to the chorus. Additionally, fed males called more often than control males, a novel response for anurans. A significantly higher calling rate was noted from males even 31 nights after supplementary feeding.

Conclusion

This strategy of allocating additional energy reserves to increasing calling rate is beneficial given the preference of female hylids for males calling at high rates and a female's ability to detect small incremental increases in calling rate.

Background

How organisms acquire resources and allocate them to the demands of maintenance, defence, repair, storage, growth and reproduction are central questions in physiological ecology [1]. After energy and nutrients have been partitioned between these major demands, further allocations can be made. For example, investing in reproduction will vary according to age and energetic status with important fitness consequences. Numerous studies have reported morphological and behavioral attributes influencing male mating success that include display rates [2], body size [3-6], body condition [7,8], lek centrality [5,9] and lek attendance [10-13]. Many of these attributes reflect how males can acquire energy and how they allocate it during the reproductive season.

Male anurans are especially interesting subjects to study how energy is partitioned for reproduction, because calling is energetically very expensive [14] and acoustic energy can easily be partitioned between call duration, calling rate, call amplitude, number of hours calling within a night, number of nights calling within a breeding season, and the number of breeding seasons in attendance.

In anurans, inter- and intrasexual selection are important determinants of mating success [15,16]. In most species of anurans with a lek mating system studied to date, male mating success is determined by the number of nights that a male is present in a breeding aggregation, i.e. chorus tenure [17]. Thus, there should be strong selection on males to increase their chorus tenure. Paradoxically, males of most anuran species spend less than 20% of the breeding season in the chorus [17,18].

A variety of hypotheses have been suggested that might explain short chorus tenure in anurans [13,19]. First, high rates of mortality within and away from the chorus are one hypothesis. Second, chorus tenure could be underestimated if males move between ponds and monitoring is restricted to a single chorus. The energetic limitation hypothesis offers a third explanation for short chorus tenure in anurans. Males may need to allocate the available energy into the maintenance of vital processes and growth with limited energy reserves left for reproduction and attracting females. Since calling is energetically very costly [14], males will not be able to maintain high rates of calling, or be present in the chorus on many nights, conditions that are necessary to attract females.

So far, studies of energy allocation in anurans have concentrated on the effects of energetic constraints on body condition and chorus tenure. Short term allocation strategies such as modulating the number of calls produced during nightly chorus attendance or varying calling rates have received less attention [20-22]. If calling rate and chorus tenure of a male are positively correlated with his reproductive success, there should be a trade-off between both parameters since males will be unlikely to be able to allocate unlimited energy to both behaviors. Males should allocate energy preferentially to the strategy that most strongly affects male mating success.

In this paper we investigate the effects of energetic constraints on breeding behavior of male European treefrogs, Hyla arborea (Linnaeus, Anura: Hylidae), a species of considerable conservation interest [23,24]. We hypothesized that if breeding behavior is energetically limited, then males should lose body condition between their first and last night in the breeding chorus. If additional energy is provided experimentally, then fed and unfed males should have different final conditions and different rates of change in condition. If the prime target of selection is to increase chorus tenure, then males provided with food should be in the chorus more nights than unfed males. In addition, fed males should return to the chorus sooner than unfed males. If the prime target of selection is to call at high rates and thus out-compete other males, then fed males should allocate the available energy to calling on few nights and should have a higher rate of calling than unfed males.

Results

Body Condition

The mean snout-vent length and tibiafibula length of captured males was 41 ± 2 mm (N = 45) and 21 ± 1 mm (N = 45), respectively. The average initial weight was 5.65 ± 0.78 g. The median initial body condition was 0.09 g for Fed1, 0.05 g for Fed2 and -0.04 g for the control males. Males in the three groups did not differ in their initial body condition (Kruskal-Wallis: H = 0.8, N = 43, p > 0.05).

No significant relationship was detectable between the first noted presence at the pond and the condition that same day (Pearson correlation: r = -0.21, N = 44, p > 0.05). However, the control males (paired t-test: T = 2.47, N = 16, p < 0.05) as well as Fed1 males (paired t-test: T = 2.81, N = 8, p < 0.05) and Fed2 males (paired t-test: T = 3.48, N = 10, p < 0.01) lost weight significantly during the season (Table 1). The average final condition was not influenced by feeding (Table 1). Fed as well as unfed males were in the same condition at the end of the season (ANOVA: F = 0.28, N = 34, p > 0.05). In addition, the supply of supplementary energy showed no influence on the rate of change in body condition (Kruskal-Wallis: H = 1.29, N = 35, p > 0.05; Table 1).

Chorus Tenure

In 2002, chorus tenure varied between the individual males. The median chorus tenure for Fed2 was 5.5 nights; for the control males and Fed1 the median chorus tenure was 7 nights (Fig. 1). Thus, both the control males and Fed1 males were present in the chorus for only 18.4% of the possible time. Fed2 spent only 14.5% of the nights in the chorus. Five males were only present 1 night (one Fed1-, two Fed2- and two control males). On average, control males stayed 19.2 nights in the calling aggregation (first until final night). This time interval amounted to 19.0 nights for Fed1 and to 12.3 nights for Fed2.

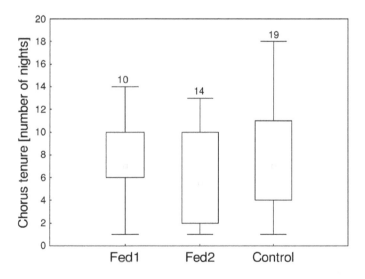

Figure 1. Male chorus tenure. Chorus tenure of males according to their treatment group (Fed1; Fed2; Control) in the year 2002 (medians, 1. and 3. quartile); the values insight the figure indicates the sample size.

There were no significant differences in chorus tenure between the treatment groups (Anova: F = 0.617, N = 42, p > 0.05). Males, which were fed, did not return to the chorus for more nights compared to males that were not provided with extra food.

Among the fed males (Fed1 and Fed2 pooled) the mass of consumed crickets (and thus the quantity of energy taken in) did not explain chorus tenure (linear regression: $r^2 = 0.054$; T = -1.1, N = 23, p > 0.05). Males that had consumed more energy did not stay in the chorus longer than males that had taken up less energy.

For fed males, the median number of nights between the treatment night and the first night that they returned was 1. Control males were absent longer (median = 2). Fed males returned back into the chorus after a significantly shorter time (Mann-Whitney U-test: $N_1 = 21$, $N_2 = 17$, U = 109, p < 0.05; Fig. 2).

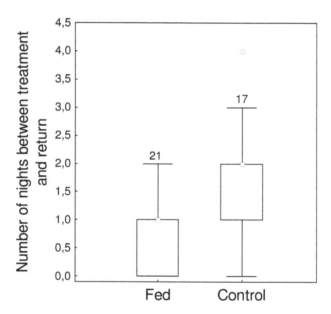

Figure 2. Number of nights between treatment and return. Number of nights between the treatment night of the males and the first night, they returned to the chorus in 2002 according to their treatment group (Fed1 + Fed2 (Fed); control); median, 1. and 3. quartile and also outlier are shown; the values insight the figure indicates the sample size.

Calling Rate

Additional available energy was invested into the rate of calling. Control males had a significantly lower calling rate than fed males (Fed1 and Fed2 pooled; paired t-test: N = 8, T = -3.1, p < 0.05; Fig. 3). This difference in calling rate was shown despite the often long delay (1 to 31 nights) between call recordings and feeding treatment. In 2002, the recordings were taken between a time period of 1 to 31 nights (median = 21 nights) after the treatment; in 2003 they were taken 1 to 2 nights (median = 1.5 nights) after the treatment.

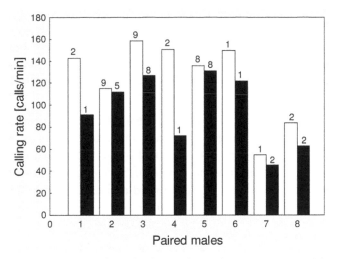

Figure 3. Calling rate. Calling rates of paired males according to their treatment group (white bars: Fed1 and Fed2 (Fed); black bars: Control); the numbers indicates the chorus tenure of the males until recordings were taken.

Control males that entered the analysis called at a rate of 95.5 ± 32.3 calls/min, whereas fed males called at 124 ± 36.9 calls/min. Fed1 males received a median of 0.078 g crickets. With an assimilation efficiency of 0.76 [25] this corresponds to a median energy input of 463.62 J. With an average weight of 5.6 g and an average oxygen consumption during calling of 1.076 ml/g·h [20] a call corresponds to 0.0157 J (conversion: 1 litre of used up O2 = 20.08 KJ [26]). With a mean calling rate of 102.2 calls/min males took up enough energy for a further 29,400 calls. Fed2 males consumed a median of five crickets corresponding to 0.25 g, whereby they received sufficient energy for an additional 94,250 calls.

Mating Success

During the reproductive season we observed six pairs in amplexus. Two fed males and two control males paired once and one fed male paired twice. Within this representative sample, the mating success of a male was positively correlated with his chorus tenure (Spearman rank correlation: $r_S = 0.51$, N = 44, p < 0.001).

Discussion

For male tree frogs Hyla arborea, the number of nights they stayed within the chorus was a crucial factor for their mating success. However, males showed an short chorus tenure [24,27,28], which is typical of most anurans [13,17,29]. At first the results seem paradoxical. If males can increase their mating success by an increased chorus tenure, why then are they absent during the largest part of the reproductive season? As a possible explanation we tested the hypothesis of energetic limitation. Under this hypothesis males would exhibit reduced chorus tenure, because they are not able to compensate the energetic costs of calling (i.e. by foraging).

As in many other anurans [14] chorus attendance was energetically expensive for male European tree frogs [20]. Males lost significant mass during the investigation period. On average, they lost 0.12 – 0.41 g between their first and last night in the chorus. With a mean weight of 5.65 g males lost between 0.12 – 0.53% of their body mass each night. But fed and unfed males did not differ in final condition or in the rate of change in condition. These results show that males do not invest the energy gained by supplementary feeding into maintaining or establishing energy reserves.

H. arborea males can reach momentary rates of oxygen consumption 41-times resting rate during calling [20]. These are the highest aerobic scopes measured so far in any ectothermic vertebrate. The high energetic cost of calling forces males to trade-off call duration, calling rate, call amplitude, number of hours calling within a night, number of nights calling within a breeding season, and the number of breeding seasons in attendance. Our study shows, that male H. arborea invest additional energy obtained from food in two ways: they return to the chorus sooner and increase their calling rate.

As in this study, similar feeding experiments of Rana catesbeiana [30], Rana clamitans [31] and Physalaemus pustulosus [32] showed no increase in male chorus tenure. In contrast, Murphy [33] found an increased chorus tenure due to supplementary feeding in Hyla gratiosa. In Hyla arborea, the energy input seems to have had a short term effect on a male's presence at the pond. Fed males returned back to the chorus after significantly fewer nights than control males suggesting that fed males were able to recover from calling activity sooner.

Calling rate is an important determinant of female choice in most anurans [15,16]. In behavioral tests, females of most anuran species that have been tested prefer males that call at high rates. This preference is generally robust even under acoustically complex field conditions (reviewed by [34]).

If males can increase their reproductive success by calling at higher rates, they should do so if they have sufficient energy reserves. Hyla arborea is the first anuran known to allocate supplementary food to calling rate. Males which were supplied with additional energy by supplementary feeding, showed significantly higher calling rates than the control males (Fig. 3). Fed males showed an average calling rate of 124 calls/min, whereas control males produced an average of 95.5 calls/

min. It is remarkable, that this difference concerning the calling rate was still detected several days after feeding. In the first year of the study, calling rate was recorded between the 1st and 31st night after males had been fed. In all cases, males seemed to allocate the additional energy in increased calling rate over a period of many nights instead of investing all the energy immediately during the first few nights following feeding.

This strategy of measured energy allocation provides males with the ability to call at higher rates than their competitors and thus remain attractive over many nights. Females of other hylids have been shown to prefer males calling at high rates [15]. A large difference in calling rate in comparison to competitors, however, does not translate proportionally to mating success. In Pseudacris crucifer [35] and Hyperolius marmoratus [36] females are able to discriminate differences in calling rate of just 12 calls/min (16% difference) and 7 calls/min (15.6% difference), respectively. Additional increases in calling rate did not further increase female preferences. Selection should therefore favour the strategy of a slightly increased calling rate (compared to any competitors), which could be kept up not only for one night but several [37].

In our study, the difference in calling rates of fed males and control males with the same total chorus tenure suggests a trade-off between calling effort and chorus tenure. In this context it would be interesting to determine if and to what degree higher calling rates are preferred by female European tree frogs.

Friedl [38] was not able to show a correlation between the mating success of Hyla arborea males and their calling rate. But his method of determining reproductive success was based on the assumption that females can choose between all males present in the chorus. This is highly unlikely and unprecedented. Most likely, females show selective attention for a subset of the males present to minimize the costs of mate sampling thus reducing predation pressure, time, energy, and opportunity costs [39-42]. Such a comparative mate sampling behavior by females is described by Friedl & Klump [43] in the course of their field observations of H. arborea, whereby the females seemed to assess only a few males before they made their mating decision. Additionally, the results of Friedl [38] could be due to male density as well as spatial and temporal pattern in their study population. Above all, if the effect is small, it may be impossible to demonstrate without big sample sizes and multiple year studies. The fact that fed males showed higher calling rates suggest that males would on average benefit from higher mating success at least over evolutionary time.

Conclusion

Our study suggests that male calling rate is an important criterion of female mate choice. Males invested the additional energy gained by feeding into increasing calling rates. Furthermore, males showed higher calling rates in larger choruses (Meuche & Grafe unpublished) probably due to competition with other males to attract females. In this context, it would be important for future investigations to determine which mate sampling tactic female European treefrogs are using as well as which and to what degree parameters such as calling rate are preferred.

Methods

General Methods

We studied Hyla arborea in southern Germany (Steigerwald) (10.6°E, 49.8°N; 480 m above sea level). Two ponds, formerly but no longer used for raising carp, served as the main study site. Other potential breeding sites were more than 5 km away. There was a distance of 30 m between the two study ponds favoured by the males. They were monitored for 2 years (from 2002 to 2003) whereas only in the year 2002 the presence of males was checked systematically. We checked on arriving and departing treefrogs by installing a drift fence that completely encircling the shore line of the smaller pond with the highest activity. The fence was a slightly modified version of the one used by Murphy [44]. Preliminary tests showed that the fence was an impenetrable barrier for Hyla arborea [28]. The fence was patrolled every 5 – 10 minutes. At the same time, the surrounding area of the first pond and especially the second pond (which had no fence) were systematically checked for calling males. With this method, it was possible to identify 71% of calling males attending the second pond each night. Here the checks on calling males started 12 days after the start of the investigation. For most males the calling activity was restricted to just one pond [28]. The breeding period lasted for 54 nights in 2004 (2nd May – 25th June). The main investigation period (14th of May – 25th of June) lasted over 43 nights, excluding five days during which no data was recorded.

A total of 47 different males were caught and marked individually. For marking, both the conventional method of toe clipping and the implantation of the VIAlpha tags (Soft Visible Implant Alphanumeric tags) of Nortwest Marine Technology Inc. (Shaw Island WA, USA). were used. These pliable, fluorescent VIAlpha tags (1.0 × 2.5 mm) were injected laterally under the frog's skin. They have a coding scheme of three alphanumeric characters and a fluorescent orange background. Detection is enhanced with UV-Light. Because the tags are made from a biocompatible medical trade elastomer they do not irritate the tissue at the implant site. VIAlpha tags were successfully tested in a number of frog species [45]. With regard to toe clipping, we only marked two toes or fingers per frog and only one toe or finger per limb. Furthermore, the first and second finger as well as the fourth toe was not marked so as to avoid impacting their climbing abilities. Neither chorus tenure nor the rate of change in condition of males varied with the marking methods [46].

Feeding Experiments

During the main investigation in 2002, departing males, on their first night at the breeding aggregation, were provided with supplementary energy in the form of crickets. If a male had not left the breeding aggregation during the first night, it was fed on the second night. In 2003, the feeding experiments were carried out on only three consecutive days. Within seasons, males were not treated more than once.

For feeding experiments, males were put in small plastic boxes (10 × 10 cm) perforated for air supply and containing wet cellulose to prevent them from dehydration. The animals were divided into three groups: 1) Fed1 – males were offered one cricket, 2) Fed2 – males were supplied with 10 crickets, not all of which were eaten, 3) control males were not fed. Males were randomly assigned to one of the experimental groups. To increase sample size, animals that did not eat any of the offered crickets were later assigned to the control group (N = 6). All individuals were kept in the containers for 12 hours. Apart from the reassigned males, control males had no opportunity to feed while they were restrained.

For every fed male the amount and mass of the consumed crickets was determined. The average amount of energy contained in a gram cricket is 8033 J [25]. With an assimilation efficiency of 0.76 [25] a male tree frog acquires 5944 J/g digested cricket [47].

Condition

If breeding behavior is energetically costly then males should lose condition between their first and their last night in the chorus, unless they can compensate by increasing energy uptake. Furthermore, males with longer chorus tenure should have a higher initial body condition, a lower final condition or a lower rate of change in condition. To test these predictions, males were measured with a vernier calliper (measurement error: ± 0.1 mm) when they were first caught and body mass taken with a portable scale (Satorius Handy; measurement error: ± 0.01 g).

Before weighing, males were put in a container with pond water to fully rehydrate them.

An Index of Condition (sensu [33,48]) was calculated for every night a male was weighed. A male's condition was determined by regression between the initial body mass of all males and their tibiafibula length. The initial condition of a male was defined as the deviation of its body mass from the predicted body mass by regression (residuals). The final condition was defined as the difference between the final body mass of a male on its last visit and the body mass calculated by regression (i.e. initial body mass to tibiafibula length). The rate of change in condition of a male was determined by the difference between the starting and the final condition divided by the number of nights between their first and last night in the breeding aggregation. All nights were counted, from the first to the final visit of a male in the chorus, irrespective of whether a male was present in the intervening period or not.

To determine if all males of each group had lost weight during their first and their final night at the breeding aggregation, only those animals were analysed that stayed at the breeding aggregation for more than one night after the feeding treatment. When measurement data of the final night were missing, the data of the last measurement was taken instead (N = 17).

Chorus Tenure

The chorus tenure of each male was recorded and compared between all groups. As the weather condition was different every night, this could have had an impact on the chorus tenure of the treated male. To avoid such seasonal effects, we paired Fed1, Fed2 and control males by night.

Recording of Calls

On several nights between 31st of May and 18th of June 2002 the rate of calling was recorded at the two study ponds. In addition, there were recordings at all four ponds between 24th of May and 14th of June 2003. Calls were recorded using several recorders (Sony Professional Walkman WM6DC and Marantz PMD 430) and microphones (Vivanco EM32) simultaneously. The recordings started after the breeding aggregation had settled and the calling sites/territories had been established. We moved slowly towards a group of calling males and recorded one or several cycles of calling. During a cycle, each male was recorded a minimum of 5 – 10 minutes. Males stopped calling only briefly when approached. Most recordings were made between 2300 and 0130. During recordings, the air and water temperature near calling males was measured by means of a digital thermometer.

The recordings were used to determine the calling rate of males. For statistical analysis of calling rate, the values of Fed1 and Fed2 males were pooled and compared to the control group using a paired design to control for variability in calling rate due to variation in chorus size (Meuche & Grafe unpublished, [49]) and temperature [49-51] factors known to strongly influence calling rate. To control the impact of these social and climatic factors, one control male and one fed male, which had been calling simultaneously, were paired. In addition, only those callers that were both calling in the same microhabitat (water or land) were compared. To avoid pseudoreplication, males were marked individually by toe-clipping [46].

Since the arrival of males was highly synchronized (Meuche & Grafe unpublished), it is unlikely that differences in calling rate between control and fed males are the result of different times of arrival. If not stated otherwise, means ± SD are given, and all tests are two-tailed.

Authors' Contributions

IM and TUG participated in design of the study and data collection. IM performed the statistical analysis and wrote the manuscript. TUG helped to draft the manuscript. All authors read and approved the final manuscript.

Note

Table 1 – Changes in weight and condition

Medians of changes in weight and condition for Fed1, Fed2 and Control (K) males during the reproduction season in the year 2002.

Acknowledgements

We thank Eduard Linsenmair, Hans-Joachim Poethke, Erhard Strohm and Minatallah Boutros for logistical support. Fieldwork was greatly aided by Johannes Penner, Renè Nestler, Laura Bollwahn, Roy Becher, Torsten Wenzel, Johannes Bitz, Frank Scheiner as well as the Meuche family and numerous students. Special thanks to Laura Bollwahn, Andreas Senkel and Steve Pike for revising an early version of the manuscript. The Animal Care Commission of Würzburg University and the Nature Conservation Office of Lower Frankonia granted permission to conduct this study.

References

1. Townsend CR, Calow P: Physiological Ecology. Sunderland: Sinauer Associates; 1981.

2. Höglund J, Lundberg A: Sexual selection in a monomorphic lek-breeding bird – correlates of male mating success in the great snipe Gallinago media. Behavioral Ecology and Sociobiology 1987, 21:211–216.

3. Elmberg J: Factors affecting male yearly mating success in the common frog, Rana temporaria. Functional Ecology 1991, 28:125–131.

4. Morrison C, Hero JM, Smith WP: Mate selection in Litoria chloris and Litoria xanthomera: females prefer smaller males. Austral Ecology 2001, 26:223–232.

5. Shorey L: Mating success on white-bearded manakin (Manacus manacus) leks: male characteristics and relatedness. Behavioral Ecology and Sociobiology 2002, 52:451–457.

6. Wells KD: Reproductive behavior and male mating success in a neotropical toad Bufo typhonius. Biotropica 1979, 11:301–307.

7. Cherry MI: Sexual selection in the raucous toad, Bufo rangeri. Animal Behavior 1993, 45:359–373.

8. Dyson ML, Henzi SP, Halliday TR, Barrett L: Success breeds success in mating male reed frogs (Hyperolius marmoratus). Proc Biol Sci 1998, 265(1404):1417–1421.

9. Hovi M, Alatalo RV, Höglund J, Lundberg A, Rintamaki PT: Lek center attracts black grouse females. Proc Roy Soc Lond B 1994, 258:303–305.

10. Bertram S, Berril M, Nol E: Male mating success and variation in chorus attendance within and among breeding seasons in the gray treefrog (Hyla versicolor). Copeia 1996, 3:729–734.

11. Fiske P, Kalas JA, Saether SA: Correlates of male mating success in the lekking great snipe (Gallinago media) – results from a four-year study. Behavioral Ecology 1994, 5:210–218.

12. Lanctot RB, Weatherhead PJ, Kempenaers B, Scribner KT: Male traits, mating tactics and reproductive success in the buff-breasted sandpiper, Tryngites subruficollis. Animal Behavior 1998, 56:419–432.

13. Murphy CG: Chorus tenure of male barking treefrogs, Hyla gratiosa. Animal Behavior 1994, 48:763–777.

14. Wells KD: The energetics of calling in frogs. In Anuran Communication. Edited by: Ryan MJ. Washington, DC: Smithsonian Institution Press; 2001:45–60.

15. Gerhardt HC, Huber F: Acoustic communication in insects and anurans. University of Chicago Press; 2002.

16. Wells KD, Schwartz JJ: The behavioral ecology of anuran communication. In Hearing and Sound Communication in Amphibians. Edited by: Narins PM, Feng AS, Fay RR, Popper AN. New York: Springer Verlag; 2007:44–86.

17. Halliday TR, Tejedo M: Intrasexual selection and alternative mating behavior. In Amphibian Biology. Social Behavior. Volume II. Edited by: Heatwole H, Sullivan BK. Surrey Beatty, Chipping Norton; 1995:419–468.

18. Murphy CG: The mating system of the barking treefrog (Hyla gratiosa). PhD thesis. Cornell University; 1992.

19. Emerson SB: Male advertisement calls: behavioral variation and physiological processes. In Anuran Communication. Edited by: Ryan MJ. Washington: Smithsonian Institution Press; 2001:36–44.

20. Grafe TU, Thein J: Energetics of calling and metabolic substrate use during prolonged exercise in the European treefrog Hyla arborea. J Comp Physiol [B] 2001, 171(1):69–76.

21. Schwartz JJ, Ressel SJ, Bevier CR: Carbohydrates and calling: Depletion of muscle glycogen and the chorusing dynamics of the neotropical treefrog Hyla microcephala. Behavioral Ecology and Sociobiology 1995, 37:125–135.

22. Wells KD, Taigen TL: Calling energetics of a neotropical treefrog, Hyla microcephala. Behavioral Ecology and Sociobiology 1989, 25:13–22.

23. Glandt D, Kronshage A: Der Laubfrosch (Hyla arborea L.) Biologie – Schutzmaßnahmen – Effizienzkontrolle. Zeitschrift für Feldherpetologie 2004, S5:63–71.

24. Pellet J, Helfer V, Yaniic G: Estimating population size in the European tree frog (Hyla arborea) using individual recognition and chorus counts. Amphibia-Reptilia 2007, 28:287–294.

25. Smith GC: Ecological energetics of three species of ectothermic vertebrates. Ecology 1976, 57:252–264.

26. Schmidt-Nielsen K: Animal physiology. Cambridge: Cambridge University Press; 1990.

27. Grafe TU, Meuche I: Erratum to Grafe, T.U. & Meuche, I. (2005). Amphibia-Reptilia 2006, 27:157.

28. Grafe TU, Meuche I: Chorus tenure and estimates of population size of male European treefrogs Hyla arborea: implications for conservation. Amphibia-Reptilia 2005, 26:437–444.

29. Given MF: Interrelationships among calling effort, growth rate and chorus tenure in Bufo fowleri. Copeia 2002, 4:979–987.

30. Judge KA, Brooks RJ: Chorus participation by male bullfrogs, Rana catesbeiana: a test of the energetic constraint hypothesis. Animal Behavior 2001, 62:849–861.

31. Gordon NM: The effect of supplemental feeding on the territorial behavior of the green frog (Rana clamitans). Amphibia-Reptilia 2004, 25:55–62.

32. Green AJ: Determinants of chorus participation and the effects of size, weight and competition on advertisment calling in the tùngara frog, Physalaemus pustulosus (Leptodactylidae). Animal Behavior 1990, 39:620–638.

33. Murphy CG: Determinants of chorus tenure in barking treefrogs (Hyla gratiosa). Behavioral Ecology and Sociobiology 1994, 34:285–294.

34. Grafe TU: Anuran choruses as communication networks. In Animal Communication Networks. Edited by: McGregor PK. Cambridge: Cambridge University Press; 2005:277–299.

35. Gerhardt HC: Female mate choice in treefrogs: static and dynamic acoustic criteria. Animal Behavior 1991, 42:615–635.

36. Grafe TU: Costs and benefits of mate choice in the lek-breeding reed frog, Hyperolius marmoratus. Animal Behavior 1997, 53:1103–1117.

37. Schwartz JJ, Buchanan BW, Gerhardt HC: Acoustic interactions among male gray treefrogs, Hyla versicolor, in a chorus setting. Behavioral Ecology and Sociobiology 2002, 53:9–19.

38. Friedl TWP: Individual male calling pattern and male mating success in the European treefrog (Hyla arborea): Is there evidence for directional or stabilizing selection on male calling behavior? Ethology 2006, 112:116–126.

39. Greenfield MD, Rand AS: Frogs have rules: selective attention algorithms regulate chorusing in Physalaemus pustulosus (Leptodactylidae). Ethology 2000, 106:331–347.

40. Murphy CG, Gerhardt HC: Mate sampling by female barking treefrogs (Hyla gratiosa). Behavioral Ecology 2002, 13:472–480.

41. Sullivan BK, Kwiatkowski MA: Courtship displays in anurans and lizards: theoretical and empirical contributions to our understanding of costs and selection on males due to female choice. Functional Ecology 2007, 21:666–675.

42. Wilkelski M, Carbone C, Bednekoff PA, Choudhury S, Tebbich S: Why is female choice not unanimous? Insights from costly mate sampling in marine iguanas. Ethology 2001, 107:623–638.

43. Friedl TWP, Klump GM: Sexual selection in the lek-breeding European treefrog (Hyla arborea): body size, chorus attendance, random mating and good genes. Animal Behavior 2005, 70:1141–1154.

44. Murphy CG: A modified drift fence for capturing treefrogs. Herpetological Review 1993, 24:143–145.

45. Buchanan A, Sun L, RS W: Using alpha numeric fluorescent tags for individual identification of amphibians. Herpetological Review 2005, 36:43–44.

46. Meuche I, Grafe TU: Nummerierte Hautimplantate – eine alternative Markierungsmethode für den Laubfrosch (Hyla arborea)? Zeitschrift für Feldherpetologie 2004, S5:153–158.

47. Murphy CG: Nightly Timing of chorusing by male barking treefrogs (Hyla gratiosa): the influence of female arrival and energy. Copeia 1999, 1999:333–347.

48. Jakob EM, Marshall SD, Uetz GW: Estimating fitness: A comparison of body condition indices. Oikos 1996, 77:61–67.

49. Friedl TWP, Klump GM: The vocal behavior of male European treefrogs (Hyla arborea): implications for inter- and intrasexual selection. Behavior 2002, 113–136:139.

50. Castellano S, Cuatto B, Rinella R, Rosso A, Giacoma C: The advertisment call of the European treefrogs (Hyla arborea): A multilevel study of variation. Ethology 2002, 108:75–89.

51. Schneider H: Rufe und Rufverhalten des Laubfrosches, Hyla arborea arborea (L.). Zeitschrift für vergleichende Physiologie 1967, 57:174–189.

CITATION

Meuche I and Grafe TU. Supplementary Feeding Affects the Breeding Behavior of Male European Treefrogs (Hyla arborea). BMC Ecology 2009, 9:1 doi:10.1186/1472-6785-9-1. © 2009 Meuche and Grafe; licensee BioMed Central Ltd. Originally published under the Creative Commons Attribution License, http://creativecommons.org/licenses/by/2.0

Mirror-Induced Behavior in the Magpie (*Pica pica*): Evidence of Self-Recognition

Helmut Prior, Ariane Schwarz and Onur Güntürkün

ABSTRACT

Comparative studies suggest that at least some bird species have evolved mental skills similar to those found in humans and apes. This is indicated by feats such as tool use, episodic-like memory, and the ability to use one's own experience in predicting the behavior of conspecifics. It is, however, not yet clear whether these skills are accompanied by an understanding of the self. In apes, self-directed behavior in response to a mirror has been taken as evidence of self-recognition. We investigated mirror-induced behavior in the magpie, a songbird species from the crow family. As in apes, some individuals behaved in front of the mirror as if they were testing behavioral contingencies. When provided with a mark, magpies showed spontaneous mark-directed behavior. Our findings provide the first evidence of mirror self-recognition in a non-mammalian species. They suggest that essential components of human self-rec-

ognition have evolved independently in different vertebrate classes with a separate evolutionary history.

Author Summary

A crucial step in the emergence of self-recognition is the understanding that one's own mirror reflection does not represent another individual but oneself. In non-human species and in children, the "mark test" has been used as an indicator of self-recognition. In these experiments, subjects are placed in front of a mirror and provided with a mark that cannot be seen directly but is visible in the mirror. Mirror self-recognition has been shown in apes and, recently, in dolphins and elephants. Although experimental evidence in nonmammalian species has been lacking, some birds from the corvid family show skill in tasks that require perspective taking, a likely prerequisite for the occurrence of mirror self-recognition. Using the mark test, we obtained evidence for mirror self-recognition in the European Magpie, Pica pica. This finding shows that elaborate cognitive skills arose independently in corvids and primates, taxonomic groups with an evolutionary history that diverged about 300 million years ago. It further proves that the neocortex is not a prerequisite for self-recognition.

Introduction

Since the pioneering work by Gallup [1], a number of studies have investigated the occurrence of mirror-induced self-directed behavior in animals of a great range of species. Most animals exposed to a mirror respond with social behavior, e.g., aggressive displays, and continue to do so during repeated testing. In a few ape species, however, behavior changes over repeated presentations with a mirror. Social behavior decreases, and the mirror is used for exploration of the own body. This suggestive evidence of self-recognition is further corroborated by the mirror and mark test. If an individual is experimentally provided with a mark that cannot be directly seen but is, however, visible in the mirror, increased exploration of the own body and self-directed actions towards the mark suggest that the mirror image is being perceived as self. Fairly clear evidence of this has been obtained for chimpanzees [1], orang-utans [2], and pygmy chimpanzees [3]. In gorillas and gibbons, some authors reported failure of self-recognition [4,5] whereas others reported positive findings in at least one individual [6,7]. It should be mentioned that even in the chimpanzee, the species most studied and with the most convincing findings, clear-cut evidence of self-recognition is not obtained in all individuals tested. Prevalence is about 75% in young adults and considerably less in young and aging individuals [8]. Findings suggestive of self-recognition in mammals other than apes have been reported for dolphins [9] and elephants [10]. In monkeys, nonprimate mammals, and in a number of bird species, exploration of the mirror and social displays were observed, but no hints at mirror-induced self-directed behavior have been obtained [5]. Does this mean a cognitive Rubicon with apes and a few other species with complex social behavior on one side and the rest of the animal kingdom on the other side? This might imply that animal self-recognition is restricted to mammals with large brains and highly evolved social cognition but absent from animals without a neocortex.

Within humans and apes, self-recognition might reflect a homologous trait, whereas findings in other mammals hint at a convergent evolution. A likely reason for such convergent evolution of self-recognition in dolphins and elephants is the convergent evolution of complex social understanding and empathetic behavior [10]. If self-recognition is linked to highly developed social understanding, some birds species, in particular from the corvid family, are likely candidates for self-recognition, too. A number of studies from the past years have demonstrated an elaborated understanding of social relations, in particular during competition for food. It has been shown that own experience in pilfering caches facilitates predicting similar behavior in others [11], and that magpies [12] and scrub jays [13] remember who of their conspecifics observed them during storing. Thus, food-storing birds might be particularly apt in empathy and perspective taking, which have been suggested to coevolve with mirror self-recognition [14].

An investigation of self-recognition in corvids is not only of interest regarding the convergent evolution of social intelligence, it is also valuable for an understanding of the general principles that govern cognitive evolution and their underlying neural mechanisms. Mammals and birds inherited the same brain components from their last common ancestor nearly 300 million years ago and have since then independently developed a relatively large forebrain pallium. However, both classes differ substantially with regard to the internal organization of their pallium, with birds lacking a laminated cortex but having developed an organization of clustered forebrain entities instead [15]. In some groups of birds and mammals, such as corvids and apes, respectively, brain to-body ratios are especially high [16], and these animals are able to generate the same complex cognitive skills [17]. This is indicated by feats such as tool use and tool manufacture [18,19], episodic-like memory [20], and the ability to use own experience in predicting the behavior of conspecifics [11]. Although it has been shown that some birds, e.g., Grey Parrots [21], use mirrors with skill in order to localize and discriminate objects, no experimental evidence of self-recognition has been obtained in birds so far.

In the present study, magpies were chosen for several reasons. They are food-storing corvids that compete with conspecifics for individually cached and memorized hoards. They thus live under ecological conditions that favor the evolution of social intelligence [12,17]. They achieve the highest level of Piagetian object permanence [22], which is also achieved by apes, but not by monkeys. In addition to showing social understanding during competition for food [12], magpies are curious and prone to approach new situations, making them ideally suited for an experiment that requires spontaneous interaction with a new and puzzling context.

Mirror behavior in animals goes through several stages. In all species tested so far, inspection of the mirror and social behavior has been observed. In species with mirror self-recognition, some of the individuals also show evidence of inspection of their own body and testing for behavioral contingencies after familiarization with the mirror. For example, they move back and forth in front of the mirror, and this might indicate that they check to which degree the mirror image is coupled to their own movement. Individuals achieving this stage often also pass the mirror and mark test.

In our experiments, we began with open mirror exploration, and then we assessed preference for the mirror and quantified mirror-induced behaviors under highly standardized conditions in a two-compartment cage with one side containing a mirror. Subsequently, we investigated spontaneous self-directed behavior in individuals provided with a mark, and finally, we carried out a series of mark tests and control tests that were designed as to ensure appropriate control and exclude the possibility that findings are due to operant conditioning. Marked individuals (cf. Figure 1) were given a small number of tests, and we applied two types of appropriate controls. The birds were either marked with a brightly colored (yellow or red) or a black (sham) mark. Handling and somesthetic input was thus identical for all marks, but the black mark was practically not visible on the black feathers of the throat. In half of the trials, a mirror was placed with the reflective surface towards the animal; in the other half of trials, the mirror was replaced by a nonreflective plate of the same size and position. Therefore, the possibility to see a colored spot on the own body by means of the mirror was the only predictor of an increase of behavioral activities towards the marked (or sham-marked) region (Figure 2) in the different experimental conditions. Each bird was tested twice in each of the conditions, resulting in eight tests per bird.

Figure 1. Magpie with Yellow Mark

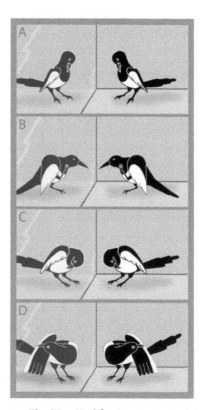

Figure 2. Examples of the Behaviors That Were Used for Quantitative Analysis (A) Attempt to reach the mark with the beak; (B) touching the mark area with the foot; (C) touching the breast region outside the marked area; (D) touching other parts of the body. Behaviors (A) and (B) entered the analysis as mark-directed behavior; behaviors (C) and (D) and similar actions towards other parts of the body were considered self-directed, but not related to the mark.

Results and Discussion

Open Mirror Test

In baseline trials with a nonreflective plate, there was no remarkable behavior in front of the plate in any of the individuals. With a mirror, the behavior of the magpies clearly differed. Initial exploration of the mirror was characterized by approaches towards the mirror and looks behind the mirror. Also, social behavior occurred, such as aggressive displays towards the mirror and jumping towards the mirror as in a fight. In three individuals, Gerti, Goldie, and Schatzi, social behaviors were transient, i.e., they were reduced already on the second exposure or completely ceased to occur. In the other two birds, Harvey and Lilly, social behaviors, in particular aggressive and submissive displays, continued to be frequent. On several trials, Harvey also picked up little, but conspicuous, objects and posed, accompanied by wing-flipping, in front of the mirror holding the objects in the beak. This courtship-like behavior vanished after a few trials, and was never seen on later tests, which were characterized by aggressive displays. In the open mirror experiment, however, mainly two of the birds (Goldie and Harvey) took part, whereas the other three birds only occasionally visited the location of the mirror. Therefore, we proceeded with a highly standardized protocol for mirror exploration.

Mirror Preference and Standardized Mirror Exploration

In these tests, birds could choose between two identical compartments of a cage, one equipped with a mirror and the other with a nonreflective plate instead of the mirror. Table 1 gives the time the birds spent with view on the mirror and shows how many bouts of close inspection of the mirror, of looks behind the mirror, of contingent behavior, and of social behavior were displayed by the birds. Three of the individuals (Gerti, Goldie, and Schatzi) spent a considerable amount of time in the compartment with the mirror, whereas the two other birds (Lilly and Harvey) appeared to avoid the compartment with the mirror. In the three birds with a preference for the mirror, behavior was characterized by close visual exploration of the mirror image. In addition, Gerti and Schatzi repeatedly looked behind the mirror and showed several bouts of behavior indicating contingency testing. Subjects moved their head or the whole body back and forth in front of the mirror in a systematic way. In Goldie, contingent behavior was not demonstrated in this test, but demonstrated later in the mark test. Harvey and Lilly never showed any hint of such behavior. It is noteworthy that those birds that had a high interest in the mirror and also showed social displays only in the first tests were those that showed at least some evidence of self-directed behavior on later tests.

Table 1. Behavioral Data from Mirror Preference and Standardized Mirror Exploration

Subject	Test	Time with View of Mirror (min:sec)	Close Inspection of Mirror Image	Looks Behind	Contingent Behavior	Social Behavior	Social Behavior Later Test	Self-Directed Mark Test	Self-Directed Other
Gerti	1	19:17	19	5	2	0	No	+	+
	2	19:24	23	5	4	0			
	3	18:21	10	2	4	0			
	4	17:12	11	0	1	0			
	5	9:26	7	1	0	0			
Goldie	1	8:54	2	0	0	3	No	+	+
	2	4:16	0	0	0	1			
	3	7:59	1	0	0	0			
	4	4:58	0	0	0	0			
	5	4:42	0	0	0	0			
Harvey	1	1:14	0	0	0	5	Yes	–	–
	2	0:06	0	0	0	0			
	3	0:27	0	0	0	0			
	4	2:05	0	0	0	0			
	5	0:00	0	0	0	0			
Lilly	1	5:47	3	2	0	4	Yes	–	–
	2	Inactive	—	—	—	—			
	3	Inactive	—	—	—	—			
	4	Inactive	—	—	—	—			
	5	Inactive	—	—	—	—			
Schatzi	1	16:05	19	8	3	4	No	(+)	+
	2	14:18	19	13	5	0			
	3	12:01	3	2	3	0			
	4	14:53	9	3	2	0			
	5	7:08	2	0	0	0			

The three birds with evidence of mirror-induced self-directed behavior on later testing spent a higher amount of time in the compartment with the mirror. Interest in the mirror tended to decline over repeated trials. All other scores are based on an event-sampling procedure. During "close inspection," the bird is very close to the mirror and inspects the image, accompanied by turning and tilting of its head. "Contingent behavior" was counted when a bird moved repeatedly leftwards and rightwards or back and forth in front of the mirror. All social behaviors displayed in these tests belonged to the agonistic context (see Materials and Methods for details and Video S9 for an example). For comparison, the three columns on the right show whether the subjects showed social behavior on later testing and whether they showed self-directed behavior in front of the mirror in the mark test or on other occasions. A negative sign (–) indicates no occurrence of the behavior, and a plus sign (+) indicates that the behavior was shown. Parentheses around the plus sign for the subject Schatzi indicate that mark-directed behavior was enhanced but did not reach statistical significance.

Mark Test

In a first exposure to a mirror with a mark, three out of five birds showed at least one instance of spontaneous self-directed behavior. In the subsequent quantitative analysis, which compared the behavior in the mirror and mark condition with a condition without a mirror, mark-directed behavior in two of the birds, Gerti and Goldie, was significantly higher in the critical mirror and mark condition than in the other conditions. There were no instances in any of the birds of pecking at the reflection of the mark in the mirror. Figure 3 shows the quantitative amount of behavior towards the mark region as a proportion of all behaviors towards the own body for these two birds. Contrary to the absolute counts, this proportion will only increase with a specific effect on mark-directed behavior but not as a consequence of a possible overall increase in behaviors towards the own body. It can be seen that mark-directed behavior was only significantly enhanced when a mirror was present and the mark was colored and thus visible for the birds. The detailed frequencies in the different conditions are given in Table 2. The comparison of the frequencies of behaviors directed towards other parts of the body clearly shows that the mark-directed behaviors in the mirror and colored mark condition cannot be explained by a general increase of behavioral activity. A specific increase of mark-directed behavior in presence of a mirror is corroborated by the fact that mark-directed behavior ceased within trials as soon as the bird had removed the mark. In Figure 4A, performance of Gerti in a single test with change of marks is shown (see also Table 3). As this is only one test, findings should be interpreted with some caution, but consistent with the between-sessions comparison in the first test series, mark-directed behavior was high when the mark was visible in the mirror and low with the black control mark that was not visible. Figure 4B shows the results of an additional test with two sessions in the mirror and colored mark (yellow) condition and two control sessions with colored marks and a nonreflective plate instead of the mirror. Also in this case, mark-directed behavior only occurred with the mirror. With colored mark and mirror, over the two trials, there were five mark-directed actions per trial and 12 actions towards the rest of the body, whereas there was no mark-directed behavior at all when the bird wore a colored mark but no mirror was present. Again, findings clearly show that this significant difference cannot be explained by an unspecific increase in overall behavioral activity, as the overall rate of behaviors directed towards the own body was similar. These quantitative data are rather conservative as Gerti and Goldie removed the mark after a few minutes on most of the trials with a color mark and a mirror, and after removal of the mark, no mark-directed behaviors occurred.

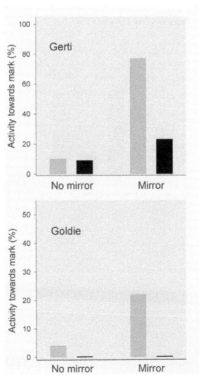

Figure 3. Proportion of Self-Directed Behavior towards the Mark Area Expressed as a Proportion of Overall Self-Directed Behavior in Subjects Gerti and Goldie. Grey bars refer to tests with a colored mark (yellow or red), black bars to tests with a black control mark (sham condition). In Gerti ($p < 0.005$, Fisher exact test), as well as in Goldie ($p < 0.05$, Fisher exact test), mark-directed behavior was significantly enhanced in the colored mark and mirror condition.

Table 2. Frequencies of Self-Directed Behaviors in the Mark Test

Subject	Test	Test with Two Trials in the Four Different Conditions			
		No Mirror/ Color	No Mirror/ Sham	Mirror/ Color	Mirror/ Sham
Gerti	1st test	0/5	0/2	9/3	3/8.5
	2nd test	1/4	1/8	5.5/1.5	1/5
	Total	1/9	1/10	14.5/4.5**	4/13.5
Goldie	1st test	1/24	0/19	4/14	0/1
	2nd test	0/0	—ᵃ	0/0	0/8
	Total	1/24	0/19	4/14*	0/9
Harvey	1st test	—ᵃ	0/0	0/3	0/4
	2nd test	—ᵇ	0/24.5	0/7	—ᵇ
	Total	—	0/24.5	0/10	0/4
Lilly	1st test	0/0	0/0	0/0	0/3
	2nd test	0/0	0/0	0/0	0/0
	Total	0/0	0/0	0/0	0/3
Schatzi	1st test	0/0	0/0	2/0	0/0
	2nd test	0/0	0/2	0/0	0/0
	Total	0/0	0/2	2/0	0/0

All individuals had two tests in four different conditions. For each entry, the first value gives the number of mark-directed actions, while the second value gives the number of self-directed actions towards other parts of the body. Differences between conditions were analyzed by comparing the total frequencies from the first and second tests with the Fisher exact test. For Gerti, the score in the mirror and color condition differs from all other conditions; double asterisks (**) indicate $p < 0.005$. For Goldie, the score in the mirror and color condition differs from the no-mirror and color as well as from the no-mirror and sham conditions; a single asterisk (*) indicates $p < 0.05$. The three instances with decimal numbers result from a slightly different rating by the two observers. For statistics, in these cases, numbers were rounded, and as such, they were conservative regarding the hypothesis.
ᵃValue missing due to loss of data during digitizing of video tapes;
ᵇTrial not completed.

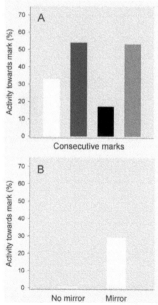

Figure 4. Proportion of Self-Directed Behavior towards the Mark Area in Additional Tests. (A) Proportion of mark-directed behavior by Gerti in a test session with change of marks in consecutive 5-min intervals. The color of each bar refers to the color of the mark used and subsequent 5-min periods of the test. Activity towards the mark was high for all colored marks, but low in the control condition (black mark). In all of the color conditions, the bird removed the mark. (B) Results from the first set of additional controls with a grid in front of the experimental cage instead of a Plexiglas wall. In the colored mark and mirror condition, mark-directed behavior was higher than in the colored mark and no-mirror condition ($p < 0.02$, Fisher exact test).

Table 3. Frequencies of Self-Directed Behaviors in the Mark Test

Test with Change of Mark Within Session in the Subject Gerti			
Mirror/Yellow	Mirror/Blue	Mirror/Sham	Mirror/Red
2/4	4/3*	0/6	11/9*

This test with Gerti with change of mark within session consisted of four consecutive 5-min periods. The asterisk (*) indicates difference from the mirror/sham condition, $p < 0.05$.

Whereas mark-directed behaviors were virtually absent when the birds were tested without a mirror but with a colored mark, there were a few instances of mark-directed behavior in the mirror condition with a black control mark (Figure 3). It may well be that the black paper mark was slightly visible on the black plumage. This is supported by the observation that if the black mark condition elicited behavior, it was in the "mirror present" trials. This is an indirect support for the interpretation that the behavior towards the mark region was elicited by seeing the own body in the mirror in conjunction with an unusual spot on the own body.

Evidence from the quantitative data is corroborated by the qualitative behavior of the birds. Self-directed activity began after looks into the mirror and visual exploration of the mirror image, and it ceased as soon as the bird had successfully removed the mark. This is unlike chimpanzees, which, after discovering that the mark is inconsequential, rapidly lose interest [23]. The reason for the difference could be that bird's feathers are of considerably higher importance for survival than a patch of hair in chimps. This interpretation is supported by data showing that birds spend about one quarter of their resting time with preening and are often seen to interrupt sleeping in complete darkness only to preen [24].

Two of the other three birds reacted to the mirror with excited behavior characterized by frequent jumping and running within the cage, and the last bird showed a high number of attacks towards the mirror in one trial, but not in the other. The subject Schatzi, which had shown spontaneous mark-directed behavior during an earlier exposure, showed no significant mark-directed behavior in this series of tests, although there were two instances of mark-directed behavior in the mirror and color condition and no mark-directed behavior in the other conditions.

Interestingly, the behavior in the mark tests corresponded to interest in the mirror in the standardized mirror exploration test. Those individuals that showed at least one instance of mark-directed behavior were the same that had shown a high interest in the mirror in the preference test, and the individual strongly avoiding the mirror in the choice test (Harvey) showed a high amount of attack-like behavior in the mark test.

One might ask why rather clear evidence was observed in two individuals and weaker evidence in another one, but not in all of the five birds. The proportion of positive findings is, however, well in the range of what has been found in apes. In chimpanzees, the species best studied and showing the clearest evidence of mirror-induced self-directed behavior, a considerable number of individuals typically produce negative findings [8,25]. Of 92 individuals tested by Povinelli et al. [8], only 21 demonstrated clear and nine weak evidence of self-exploration, with about 75% prevalence in young adults of 8 to 15 y. Only half among those with clear evidence of self-exploration passed the mark test. Thus, our data do not only qualitatively, but also quantitatively, match the findings in chimpanzees. As a note of caution, we would like, however, to emphasize that the number of birds we tested is too small for a definitive estimate of the distribution within the population. Thus, further studies must assess whether the typical frequency of mirror-induced self-directed behavior in magpies is comparable to that in chimpanzees.

Altogether, results show that magpies are capable of understanding that a mirror image belongs to their own body. We do not claim that the findings demonstrate a level of self-consciousness or self-reflection typical of humans. The findings do however show that magpies respond in the mirror and mark test in a manner so far only clearly found in apes, and, at least suggestively, in dolphins and elephants. This is a remarkable capability that is at least a prerequisite of self-recognition and might play a role in perspective taking. It thus could be essential for the ability of using own experience to predict the behavior of conspecifics [11]. Magpies are corvids, which belong to the order of Passeriformes, a phylogenetic group characterized by large brains relative to body weight [26]. The relative brain size of passeriform birds is similar to primates in allometric analyses, and within the Passeriformes, corvids stand out with particular high relative brain size [27]. Thus, magpies belong to a group of animals with very high relative brain size (see also Table 4).

We used a small number of tests as it was crucial to ensure that possible self-directed behavior of the birds represented a spontaneous response to seeing the own body in the mirror. Epstein et al. [28] reported that prolonged operant conditioning of isolated components of the mark test in pigeons could produce a behavioral pattern that superficially looks like mirror-induced mark-directed behavior. This study could, however, not be replicated [29], and these authors also found reduction of self-directed behavior in pigeons in the conditions with a mirror, which strikingly contrasts with our findings in magpies. Additionally, extremely long periods of exposure to mirrors without specific training of self-related actions did not produce any kinds of behavior that was centered on the mark in monkeys

[30,31]. Lastly, the mark test is only one piece of evidence of mirror-induced self-recognition in animals. Of equal importance are previous inspections of the mirror, such as during looks at the back side of the mirror, and exploration of mirror properties, such as during contingent behavior [1,8,10,32]. Our magpies showed self-related behavior in front of the mirror after a rather short cumulative exposure time and without being specifically trained to do so. In addition, when confronted with mirrors the first time, they displayed similar sequences of behavior as described in apes [1,2,3,8,32]. Although the mark test has been criticized [33,34], the main objections have been ruled out [23], and it remains one of the most useful tests for self-recognition in comparative studies [35]. When magpies are judged by the same criteria as primates, they show self-recognition and are on our side of the "cognitive Rubicon." One should keep in mind that though mirror self-recognition reflects a crucial step in the emergence of self-recognition, the fully fledged capacity is complex, and comparative [36,37], clinical [38], and developmental studies [39] suggest an overall gradual development of this capacity.

Table 4. Brain Weights, Body Weights, as Well as Percent Brain-to-Body Weights of Some of the Different Species That Have Been Tested with Mirrors

Species	Brain Weight (g)	Body Weight (kg)	(% Brain Weight) × 1,000
European Magpie	5.8	0.19	31
African Grey Parrot	9.18	0.405	22.6
Pigeon	2.4	0.5	5
Human	1,350	65	21
Chimpanzee	440	52	8
Gorilla	406	207	2
Rhesus Monkey	68	6.6	1
Asian Elephant	7,500	4,700,000	1.6
Bottlenose Dolphin	1,600	170	9
Cat	25.6	3.3	8

Since the percent brain-weight measure favors small species, an alternative depiction would be the encephalization quotient (EQ). The EQ indicates the extent to which the brain weight of a given species deviates from the expected brain weight based on a standard species of the same taxon or a common allometric regression. However, equations for allometric lines vary among studies, and different data are often used in different studies. To our knowledge, EQs for the species in Table 4 were never calculated with the same allometric regression.

Cognitive and neurobiological studies of the last decade have shown that birds and mammals faced a similar selection pressure for complex cognitive abilities, resulting in the evolution of a comparable neural architecture of their forebrain association areas [40] as well as their cognitive operations [17,41–43]. This high degree of evolutionary convergence is especially visible for the cognitive abilities of corvids and apes [17]. By demonstrating self-recognition in the mirror by magpies, the present study shows that even the neural capacity for distinguishing self and others has evolved independently in the two vertebrate classes and that a laminated cortex is not a prerequisite for self-recognition.

Materials and Methods

Five adult hand-raised magpies served as subjects throughout the study. These birds had been used before in a study on the development of object permanence and in patterned string problems. Investigation of behavior towards a mirror consisted of three steps.

Open Mirror Exploration

In a 4 × 4-m room, a mirror 55-cm wide and 40-cm high was placed on the ground, leaning against a pole, and slightly tilted. The position of the mirror was in the middle of one of the walls with about 1.5-m distance from the wall, allowing the birds to move around the mirror. The tested subject could move freely in the room. After a baseline trial with the mirror replaced by a grey, nonreflective plate, five test sessions of 30-min duration were given to each of the birds. The behavior of the birds was observed from an adjacent room by means of a video system, and trials were videotaped.

Mirror Preference and Standardized Mirror Exploration

For a quantitative estimation of the interest in the mirror, a cage with two opposite compartments was used. Compartments 60 × 100 × 60 cm (length × height × width) were identical except that there was a mirror at the end of one of the compartments and a grey, nonreflective plate of the same size in the other compartment. Between the compartments, there was a partition with two overlapping walls so that the birds could move freely between compartments but could not see from one compartment into the other. Each of the birds received five consecutive trials of 20 min on separate days. The time the birds spent in the compartment with the mirror was measured, and based of the videotapes bouts of close mirror inspection, looks behind the mirror, bouts of contingent behavior, and social behaviors were counted.

Mark Test

In the mark test, each subject was involved in eight test sessions with the conditions, (1) mirror and colored mark, (2) mirror and black mark, (3) no mirror with colored mark, and (4) no mirror with black mark. One session lasted 20 min. Each condition was replicated once, and two colors, either yellow or red, were used for the colored mark. Thus, our subjects were marked in all conditions to prevent cueing by somesthetic input, but the black mark in condition 2 was practically not visible on the black feathers of the throat. Such sham marking, which also had been used in studies with dolphins and elephants [9,10], has the advantage that no anesthesia is needed, and it provides a rather rigorous control as except for the difference in appearance, every detail of the procedure is perfectly matched to the conditions with a color mark. In the no-mirror controls, the mirror was replaced by a nonreflective flat grey plate of the same size and in the same position. Conditions, including usage of the two marking colors yellow and red, alternated in balanced order. Colored marks and black control marks were fixed below the beak onto the throat region (Figure 1). Different pigeon breeds are blind to this area even during strong convergent eye movements [44]. In magpies, the visual field has not been studied in detail. However, the position of the eyes and the optical axis is comparable to that of pigeons. Thus, the assumption is justified that the spot used for marking was far outside the magpies' visual field. This is, furthermore, strongly corroborated by the behavioral data, as there was virtually no mark-directed behavior if magpies wore a colored mark in the no-mirror condition. In each of the conditions, every detail regarding the handling of the birds was identical, except that the birds could see the reflection of the yellow or red mark in the mirror and colored mark condition, but not in the various other conditions.

Scoring of Behaviors

Before testing, we protocolled all behaviors observable. Then two observers (others than those assessing the mirror test) independently scored videotapes several times, and a list emerged of 18 behaviors with high interobserver agreement and high reliability over repeated scorings. The list included behaviors not relevant for the question of this study (like moving slowly or fast or jumping in the cage), and for the present study, we focused on a subset of behaviors diagnostic of the animals responding to the mirror. First, social behaviors that could be observed with high reliability were agonistic displays, either submissive or attack-like. During submissive behavior, the bird faces the mirror (or another bird), has its back lowered, its wings slightly spread, and often flips its wings. During aggressive displays, the bird takes an upright position with elongated neck and/or performs attack-like behaviors towards the mirror (c.f. [45] for pictures of typical displays by magpies). Secondly, we recorded all behaviors directed towards a bird's own body, such as touching any part of the body with the beak or the foot, and we assessed whether the action was directed towards the regions near the mark or elsewhere.

Before the first mark test, birds were familiarized with the experimental cage and the mirror for at least 5 d. The experimental cage was 120 × 100 × 60 cm (l × h × w) and had a grid floor. The walls also consisted of a grid except for one long wall, which in most of the tests consisted of Plexiglas in order to provide a good view of the bird. The mirror or the nonreflective plate was always placed on the ground at the same short wall of the cage.

At the beginning of a test, a bird was taken from its home cage and brought to an adjacent room, where the colored mark or the control mark was fixed. The bird was held by one of the experimenters such that the throat region below the beak was exposed. The head of the bird was shielded by the hand of the experimenter holding the bird so that the bird could not see the fixing procedure. The other experimenter then fixed the self-adhesive colored or black mark. Except for the color of the mark, the handling procedure was exactly the same in each of the experimental conditions. Although the dots used for marking were self-adhesive, we prepared them with double-sided adhesive tape in order to ensure good fixation. The weight of a dot was 16 µg and the diameter 8 mm.

After completion of testing of all birds in each of the conditions twice, further tests were applied for the subject Gerti, who showed very clear and consistent self-directed behavior. First of all, a mirror test with four consecutive phases of 5 min was carried out, and in each phase, a new mark was fixed (yellow, blue, black, or red). Secondly, additional tests comparing the behavior in the colored mark and mirror condition with that in the colored mark and no-mirror condition were carried out with a normal cage grid instead of a Plexiglas wall in front of the cage. In experiments with dolphins [9], the interpretation of the subjects' behaviors was complicated by the fact that the animals had apparently used the reflecting sidewalls of their pool in addition to the mirror provided by the experimenters. Although reflectance by the Plexiglas front was not likely, we sought additional controls to ensure that the presence or absence of the mirror was the only predictor of differences in mark-directed behavior.

While the birds were being set into or removed from the experimental cage, room lights were switched off. A test began by turning the lights on. Experiments were monitored via a video system from an adjacent room, and all tests were videotaped. From videotapes, the frequency of behaviors directed towards the own body were scored. Actions towards the marked region with the beak or the foot entered the analysis as mark-directed behavior (see Figure 2); all other behaviors towards the own body, such as touching the breast region with the beak, preening of the tail or wings, were considered not mark related. Likewise, all social behaviors, such as aggressive displays in front of the mirror, were scored. Figure 2 shows examples of the behaviors used for quantitative analysis. Quantitative assessment of behavior based on the video tapes was carried out independently by two observers (H. P. and O. G.). Their scores were highly correlated (r = 0.98), and the combined score from both assessments was used for further analysis.

Authors' Contributions

HP and OG conceived and designed the experiments. HP and AS performed the experiments. HP, AS, and OG analyzed the data, contributed reagents/materials/ analysis tools, and wrote the paper.

References

1. Gallup GG Jr (1970) Chimpanzees: self-recognition. Science 167: 86–87.

2. Lethmate J, Dücker G (1973) Untersuchungen zum Selbsterkennen im Spiegel bei Orang-Utans und einigen anderen Affenarten. Z Tierpsychol 33: 248–269.

3. Walraven V, van Elsacker L, Verheyen R (1995) Reactions of a group of pygmy chimpanzees (Pan paniscus) to their mirror images: evidence of self-recognition. Primates 36: 145–150.

4. Ledbetter DH, Basen JA (1982) Failure to demonstrate self-recognition in gorillas. Am J Primatol 2: 307–310.

5. Hyatt CW (1998) Responses of gibbons (Hylobates lar) to their mirror images. Am J Primatol 45: 307–311.

6. Patterson FGP, Cohn RH (1994) Self-recognition and self-awareness in lowland gorillas. In: Parker ST, Mitchell RW, editors. Self-awareness in animals and humans: developmental perspectives. New York (New York): Cambridge University Press. pp. 273–290.

7. Ujhelyi M, Merker B, Buk P, Geissmann T (2000) Observations on the behavior of gibbons (Hylobates leucogeny, H. gabriellae, and H. lar) in the presence of mirrors. J Comp Psychol 114: 253–262.

8. Povinelli DJ, Rulf AB, Landau KR, Bierschwale DT (1993) Self-recognition in chimpanzees (Pan troglodytes): distribution, ontogeny, and patterns of emergence. J Comp Psychol 107: 347–372.

9. Reiss D, Marino L (2001) Mirror self-recognition in the bottlenose dolphin: a case of cognitive convergence. Proc Natl Acad Sci USA 98: 5937–5942.

10. Plotnik JM, de Waal FBM, Reiss D (2006) Self-recognition in an Asian elephant. Proc Natl Acad Sci USA 103: 17053–17057.

11. Emery MJ, Clayton NS (2001) Effects of experience and social context on prospective caching strategies by scrub jays. Nature 414: 443–446.

12. Prior H, Gonzalez-Platta N, Güntürkün O (2004) Personalized memories for food-hoards in Magpies. Ravens Today: Third International Symposium on the Raven (Corvus corax). Metelen, Germany.

13. Dally JM, Emery NJ, Clayton NS (2006) Food-caching western scrub-jays keep track of who was watching when. Science 312: 1662–1665.

14. De Waal FBM (2008) Putting the altruism back into altruism: the evolution of empathy. Annu Rev Psychol 59: 279–300.

15. Jarvis ED, Güntürkün O, Bruce L, Csillag A, Karten H, et al. (2005) Avian brains and a new understanding of vertebrate brain evolution. Nat Rev Neurosci 6: 151–159.

16. Rehkämper G, Frahm HD, Zilles K (1991) Quantitative development of brain and brain structures in birds (Galliformes and Passeriformes) compared to that in mammals (Insectivores and Primates). Brain Behav Evol 37: 125–143.

17. Emery NJ, Clayton NS (2004) The mentality of crows: convergent evolution of intelligence in corvids and apes. Science 306: 1903–1907.

18. Hunt GR, Corballis MC, Gray RD (2001) Animal behavior: laterality in tool manufacture by crows. Nature 414: 707.

19. Weir AA, Chappell J, Kacelnik A (2002) Shaping of hooks in New Caledonian crows. Science 297: 981.

20. Clayton NS, Dickinson A (1998) Episodic-like memory during cache recovery by scrub jays. Nature 395: 272–274.

21. Pepperberg IM, Garcia SE, Jackson EC, Marconi S (1995) Mirror use by African grey parrots (Psittacus erithacus). J Comp Psychol 109: 189–195.

22. Pollok B, Prior H, Güntürkün O (2000) Development of object permanence in food-storing magpies (Pica pica). J Comp Psychol 114: 148–157.

23. Povinelli DJ, Gallup GG Jr, Eddy TJ, Bierschwale DT, Engstrom MC, et al. (1997) Chimpanzees recognize themselves in mirrors. Anim Behav 53: 1083–1088.

24. Delius JD (1988) Preening and associated comfort behavior in birds. Ann N Y Acad Sci 525: 40–55.

25. Swartz KB, Evans S (1991) Not all chimpanzees (Pan troglodytes) show self-recognition. Primates 32: 483–496.

26. Cnotka J, Güntürkün O, Rehkämper G, Gray RD, Hunt GR (2008) Extraordinary large brains in tool-using New Caledonian crows (Corvus moneduloides). Neurosci Lett 433: 241–245.

27. Iwaniuk AN, Dean KM, Nelson JE (2005) Interspecific allometry of the brain and brain regions in parrots (Psittaciformes): comparisons with other birds and primates. Brain Behav Evol 65: 40–59.

28. Epstein R, Lanza RP, Skinner BF (1981) "Self-awareness" in the pigeon. Science 212: 695–696.

29. Thompson RKR, Contie CL (1994) Further reflections on mirror-usage by pigeons: lessons from Winnie the Pooh and Pinocchio too. In: Parker S, Boccia M, Mitchell R, editors. Self-awareness in animals and humans. New York (New York): Cambridge University Press. pp. 392–409.

30. Gallup GG Jr (1977) Absence of self-recognition in a monkey (Macaca fascicularis) following prolonged exposure to a mirror. Dev Psychobiol 10: 281–284.

31. Suarez SD, Gallup GG Jr (1986) Social responding to mirrors in rhesus macaques (Macaca mulatta): Effects of changing mirror location. Am J Primatol 11: 239–244.

32. Gallup GG Jr, Povinelli DJ, Suarez SD, Anderson JR, Lethmate J, et al. (1995) Further reflections on self-recognition in primates. Anim Behav 50: 1525–1532.

33. Heyes CM (1994) Reflections on self-recognition in primates. Anim Behav 47: 909–919.

34. Heyes CM (1998) Theory of mind in nonhuman primates. Behav Brain Sci 21: 101–114.

35. Platek SM, Levin SL (2004) Monkeys, mirrors, mark tests, and minds. Trends Ecol Evol 19: 406–407.

36. De Waal FBM (2005) The monkey in the mirror: hardly a stranger. Proc Natl Acad Sci USA 102: 11140–11147.

37. Toda K, Watanabe S (2008) Discrimination of moving video images of self by pigeons (Columba livia). Anim Cogn. E-pub ahead of print. doi:10.1007/s10071-008-0161-4.

38. Feinberg TE, Keenan JE (2005) Where in the brain is the self. Conscious Cogn 14: 661–678.

39. Rochat P (2003) Five levels of self-recognition as they unfold early in life. Conscious Cogn 12: 717–731.

40. Güntürkün O (2005) The avian 'prefrontal cortex' and cognition. Curr Opin Neurobiol 15: 686–693.

41. Clayton NS, Emery NJ (2005) Corvid cognition. Curr Biol 15: R80–81.

42. Bugnyar T, Heinrich B (2005) Ravens, Corvus corax, differentiate between knowledgeable and ignorant competitors. Proc Biol Sci 272: 1641–1646.

43. Butler AB, Manger PR, Lindahl BIB, Arhem P (2005) Evolution of the neural basis of consciousness: a bird-mammal comparison. Bioessays 27: 923–926.

44. Jahnke HJ (1984) Binocular visual field differences among various breeds of pigeons. Bird Behav 5: 96–102.

45. Birkhead T (1991) The magpies: the ecology and behavior of black-billed and yellow-billed magpies. London: Poyser. 270 p.

CITATION

Precocious Locomotor Behavior Begins in the Egg: Development of Leg Muscle Patterns for Stepping in the Chick

Young U. Ryu and Nina S. Bradley

ABSTRACT

Background

The chicken is capable of adaptive locomotor behavior within hours after hatching, yet little is known of the processes leading to this precocious skill. During the final week of incubation, chick embryos produce distinct repetitive limb movements that until recently had not been investigated. In this study we examined the leg muscle patterns at 3 time points as development of these spontaneous movements unfolds to determine if they exhibit attributes of locomotion reported in hatchlings. We also sought to determine whether the deeply flexed posture and movement constraint imposed by the shell wall modulate the muscle patterns.

Methodology/Principal Findings

Synchronized electromyograms for leg muscles, force and video were recorded continuously from embryos while in their naturally flexed posture at embryonic day (E) 15, E18 and E20. We tested for effects of leg posture and constraint by removing shell wall anterior to the foot. Results indicated that by E18, burst onset time distinguished leg muscle synergists from antagonists across a 10-fold range in burst frequencies (1–10 Hz), and knee extensors from ankle extensors in patterns comparable to locomotion at hatching. However, burst durations did not scale with step cycle duration in any of the muscles recorded. Despite substantially larger leg movements after shell removal, the knee extensor was the only muscle to vary its activity, and extensor muscles often failed to participate. To further clarify if the repetitive movements are likely locomotor-related, we examined bilateral coordination of ankle muscles during repetitive movements at E20. In all cases ankle muscles exhibited a bias for left/right alternation.

Conclusions/Significance

Collectively, the findings lead us to conclude that the repetitive leg movements in late stage embryos are locomotor-related and a fundamental link in the establishment of precocious locomotor skill. The potential importance of differences between embryonic and posthatching locomotion is discussed.

Introduction

Chicks emerge from the egg after 21 days of incubation equipped to walk, swim and airstep [1]–[3]. Chicks begin moving embryonic day (E) 3. The movement (e.g., embryonic motility) is spontaneously generated throughout embryogenesis until hatching [4]–[6], which appears to be the only sensory-triggered behavior during embryonic development [5], [7], [8]. Embryonic motility is episodically generated by a recurrently connected excitatory network within the spinal cord that is transiently silenced by activity-dependent depression [9], [10]. Motility is initially driven by acetylcholine, then glutamate by E8-E10 [11]. The temporal features of the activity play an instructive role in motor neuron pathfinding [12], [13], and possibly the flexor-extensor and interlimb alternations for stepping [14]. However the relationship between the early network for motility and the locomotor network is uncertain [15], [16].

Leg movements during motility at E9 are characterized by alternating flexion and extension muscle synergies [17], [18] and joint excursions [19], [20]. Yet, electromyographic (EMG) and kinematic patterns appear to break down between E12 and E15 [17], [19], [21]. Recent studies indicate that a distinctly different pattern of repetitive leg movements (RLMs) emerges between E15 and E18 [22], [23]. RLMs are also composed of alternating flexion and extension, but the frequency range (1–10 Hz) far exceeds that for E9 motility (0.2–2 Hz). Interestingly, the RLM frequency range is similar to the combined ranges for three locomotor forms in hatchlings: walking, swimming and airstepping [2]. Together these findings raise the possibility that RLMs are an essential link between early motility and locomotion.

Leg muscles also express membership in either the flexor synergy for leg protraction (swing) or the extensor synergy for leg retraction (stance) during walking, swimming and airstepping [2]. Variations in muscle burst patterns across these behaviors appear to be due to differences in limb loading. For example, extensor burst duration exhibits a close association with step cycle duration during stance (walking), but a weak association during buoyancy (swimming) and limb suspension (airstepping). Differences in knee extensor activity also distinguish the three locomotor patterns from one another. During walking there are two knee extensor bursts; one in the latter part of swing to extend the knee and advance the foot, and one in late stance to propel the body forward [1]. Only the duration of the knee extensor burst in the stance phase co-varies with step cycle duration.

We reasoned that locomotor pattern generation must be established prior to hatching, given that chicks walk within hours afterwards. We also reasoned that RLMs might employ locomotor patterns given the similarities in cycle frequency range. However, in our previous study we observed that RLMs exhibited considerable variability in EMG activity [23]. This variability obscured ready identification of a fundamental pattern resembling any of the locomotor patterns observed in hatchlings. Thus, in this study we extend earlier findings by reporting quantitative analyses for the burst patterns at 3 ages. One aim was to determine if the EMG patterns for RLMs resemble any of the three locomotor patterns in hatchings. In addition, our earlier kinematic study of motility between E9 and E18 suggested that mechanical constraints due to growth of the body within a fixed egg volume increasingly alters joint kinematics between E15 and E18 [22]. Thus, differences we might find between locomotor and RLM muscle patterns could be attributable to constraints that include placement of flexor muscles at their shortest length and extensors at their greatest length. Therefore, another aim was to determine if EMG patterns would appear more similar to locomotor patterns when the shell constraint was removed. We provide evidence that EMG patterns for RLMs share some features common to locomotor behaviors at hatching, and the first evidence of alternating interlimb stepping in the embryo. We also provide evidence that the effects of shell removal were limited to the primary knee extensor, the only muscle that is known to distinctly vary its participation across the three locomotor behaviors in hatchlings [2].

Results

We report EMG analyses for intralimb coordination based on 1206 RLM sequences. These data represent control conditions at E18 (386 RLM, 11 embryos) and E20 (353 RLM, 12 embryos), and the experimental condition, foot-free, at E20 (467 RLM, 9 embryos). The sample was drawn from a larger sample of 1569 RLMs whose rhythm properties were previously reported [23]. For these analyses, we excluded RLMs in which the tibialis anterior (TA), an ankle flexor and the reference muscle for our analyses, was the only rhythmically active muscle or if there were fewer than 10 rhythmically stable RLMs representing an experiment. EMG patterns at E15 are also reported (10 embryos).

Leg Muscle Participation Varied within and between Experiments

The wide array of RLM rhythm frequencies and combinations of active muscles reported in our previous study raised the possibility that RLMs are a collection of rhythmic behaviors. Kinematic differences between RLMs, such as ankle motions that were in phase with proximal joints (Figure 1A) or out phase (Figures 1B, 3A) might be evidence of different limb behaviors. Quantitative analyses for this study extended those results revealing that the combination of active muscles varied markedly throughout every experiment, even in RLMs only seconds to minutes apart, though conditions were seemingly unchanged. In one E20 experiment for example, 4 leg muscles participated during one RLM (Figure 1A). Approximately 20 min later, a rhythmic sequence of TA bursts was spontaneously initiated that lacked rhythmic activity in the other EMG channels. It was followed immediately by a sequence of 6 TA cycles that included bursting in the sartorius (SA), a hip flexor (Figure 1B). In other RLMs during this experiment, TA sequences were accompanied by bursts in the femorotibialis (FT), a knee extensor (Figure 1C). To determine if there were any consistencies in recruitment, we determined the participation rate for each muscle across RLM cycles for each experiment. Based on within-subject averages (Figure 1D), the SA participated in nearly half of all RLM cycles at E18 (48%) and E20 (52%), FT participated in more than a third (38–40%), and the lateral gastrocnemius (LG), an ankle extensor, was least likely to participate at both E18 (26%) and E20 (13%). However, rates of participation for each muscle varied widely across experiments at both ages (Figure 1D).

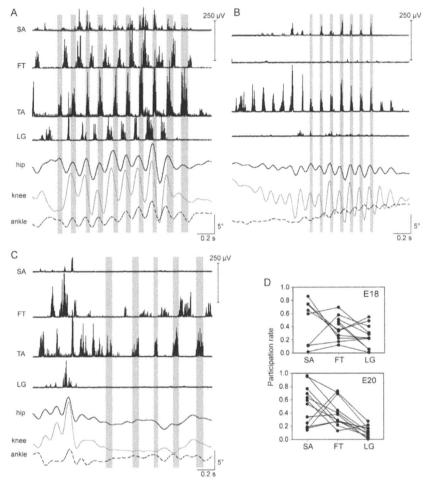

Figure 1. Variability in muscle participation during RLMs. RLMs from an E20 experiment are shown (A–C). A: This RLM consisted of 10 TA bursts and 9 rhythmically stable cycles (6.8±0.9 Hz). Vertical bars identify TA burst durations. Flexors (TA and SA) were coactive and reciprocally active with extensors (FT and LG). FT burst onsets preceded LG burst onsets. Flexors were active during flexion (downward deflections) of all 3 leg joints and extensors were active during extension (upward). B: Only TA and SA were active during 6 rhythmically stable RLM cycles (8.9±0.2 Hz). These cycles were preceded by several rhythmic TA bursts that did not meet criteria for either burst duration or rhythm stability (see Methods). Ankle excursions were minimal and opposite in direction of hip and knee. C: Only TA and FT were rhythmically active during 4 RLM cycles (4.1±0.7 Hz). Joint excursions were small and variable. D: The incidence of SA, FT, and LG bursts during RLMs is plotted for E18 (N = 11) and E20 (N = 12) embryos. The number of bursts detected by analysis methods was normalized to the total number of TA cycles per experiment (participation rate); 1 = always participated, 0 = never participated. Lines connect rates for each of the 3 muscles per experiment. SA was not implanted in 3 experiments (E18) and the LG implant was lost in 1 (E20). Abbreviations: SA, sartorius; TA, tibialis anterior; FT, femorotibialis; LG, lateral gastrocnemius.

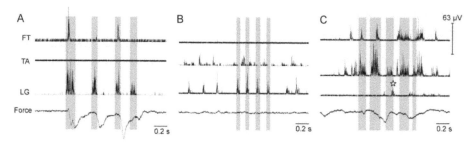

Figure 2. RLMs at E15. RLMs for 3 experiments are shown. A force transducer resting on the hip monitored general body displacements during recordings. Kinematic methods were not applied in these experiments. A: LG was rhythmically active for 3 cycles (3.2±1.0 Hz) and FT was coactive with LG. B: LG was rhythmically active for 3 cycles (6.4±0.5 Hz) and TA was reciprocally active with LG. C: TA was rhythmically active for 4 cycles (4.8±0.6 Hz), FT was coactive, and an LG burst was also detected by analysis methods (✶).

RLM sequences were difficult to detect at E15 due to the low amplitude and irregular bursting that dominated recordings. Based on a stable burst rhythm in one muscle, though not necessarily TA, 160 EMG sequences (N = 10 embryos) were examined for muscle patterns. Participation of a second muscle was detected in 24 RLMs (15%). In these instances FT bursts occurred synchronously with repetitive LG bursting (Figure 2A); TA bursts alternated with LG (Figure 2B); or FT bursts occurred synchronously with TA (Figure 2C). Participation of a third muscle was observed only twice (Figure 2C; see also [23] Figure 1F). Thus at E15, rhythmic RLM bursting was more readily apparent than any RLM muscle pattern.

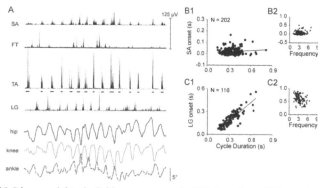

Figure 3. SA and LG bursts exhibited reliable onset patterns by E18. Data for an E18 experiment are shown. A: This RLM was 6.5 s long. Dashes underscore 15 TA bursts (2.4±0.5 Hz). TA amplitude was reduced 50% in this record to clearly visualize FT and SA traces. B–C: Larger scatter plots include trend lines for regression analyses. The number of bursts (N) in each trend analysis is indicated. In the small plots offset to the right, SA and LG burst onsets are normalized to the concurrent TA cycle duration (0–1) and plotted against TA cycle frequency (Hz). B1: SA burst onsets clustered near 0 s (slope = 0.03, R2 = 0.02). B2: SA relative onsets clustered early in the cycle (0–0.2) at all frequencies. C1: LG onset latency strongly co-varied with cycle duration (slope = 0.78, R2 = 0.76). C2: LG bursts clustered in the latter half of TA cycles with 82.8% of bursts falling between relative onsets of 0.5 and 1.0.

RLM Muscle Patterns also Exhibited Reliable Features

Though participation rates varied substantially across RLMs in an experiment, burst onsets for participating muscles were distinct. SA bursts began nearly synchronous with TA burst onset at both slower and faster RLM frequencies at both E18 and E20, with a mean onset latency of ±30 ms. An exemplary RLM and analyses for one E18 experiment are shown in Figure 3. The regression plot for SA burst onsets indicated SA onset latency clustered around 0 ms, the slope near 0.0 (Figure 3B1). Relative onsets fell mostly between 0–0.2 across a TA frequency range of 1–7 Hz (Figure 3B2). Slopes approximating 0.0 were found in 13 of 16 experiments at E18 and E20 combined.

When LG activity was well-formed, it alternated with TA bursts (Figure 3A). For example, regression results for the same E18 experiment indicated LG burst onsets co-varied closely with TA cycle duration (Figure 3C1). Relative onsets fell mostly in the latter half of the TA cycle at all RLM frequencies (Figure 3C2). These results were typical at both E18 and E20. LG onset closely varied with cycle duration in 6 of 9 experiments (R2>0.6), and was weak in only 2 (R2<0.4). Also, more than 60% of LG bursts began in the latter half of the TA cycle in 5 experiments (e.g., relative onset of 0.5–1.0), and no bias was found in 4 experiments.

The knee extensor, FT, also alternated with TA, but onset trends were distinct from LG. FT bursts exhibited 3 onset patterns relative to the TA cycle: early (relative onset <0.5), late (relative onset >0.5), or double bursting (early and late) at both E18 and E20. All 3 patterns were found within single experiments, as illustrated by 2 RLMs in Figure 4A. During the 1st RLM (Figure 4A1) an FT burst began in the first half of each cycle, and included a double burst in cycle 3 (✳). During the 2nd RLM approximately 30 s later (Figure 4A2), FT bursts began in the latter half of cycles 2–4. FT burst onsets for this experiment modestly co-varied with TA cycle duration, and 2 clusters, early and late, were observed in the relative onset plot (Figure 4A3, bottom plot). FT burst onset varied closely with cycle duration in only 7 of 17 experiments (R2>0.6). FT onset varied weakly in 8 experiments (R2<0.4), as exemplified by the E20 experiment in Figure 4B. In the latter experiment, FT bursts began early in cycles 2–6 (Figure 4B1), and during the experiment relative onsets fell mostly between 0.2–0.5 over a frequency range of 2–8 Hz (Figure 4B2, bottom plot). More than 60% of FT bursts began in the first half of the TA cycle in 10 of 17 experiments, and a bias for the latter half of the cycle was found in only 4 experiments. FT findings did not differ between E18 and E20.

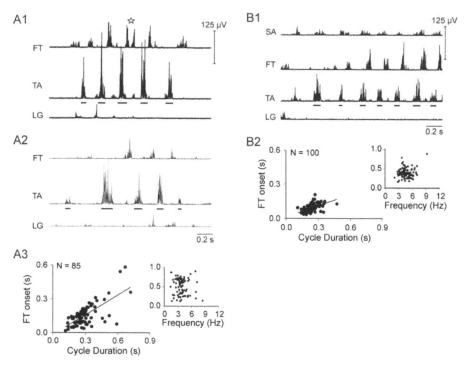

Figure 4. FT onset exhibited multiple trends. Data for an E18 (A) and E20 experiment (B) are shown. A1: FT bursts began in the first half of 4 TA cycles (3.9±0.5 Hz), accompanied by a 2nd burst in cycle 3 (✳). A2: In the next RLM 30 s later, FT bursts began in the latter half of TA cycles 2–4 (3.0±0.8 Hz). A3: FT onset and cycle duration (top) moderately co-varied (slope = 0.59, R2 = 0.48). Relative onset (bottom) was broadly distributed at all frequencies. 55% of FT bursts began in the latter half of the TA cycle. B1: FT bursts began in the first half of cycles 2–6 (4.2±0.6 Hz). B2: FT onset weakly varied with cycle duration (slope = 0.33, R2 = 0.33) and 84% of FT bursts began in the first half of the TA cycle.

Burst durations were brief for all muscles at all TA burst frequencies. Average TA burst durations slightly exceeded 60 ms and averages for SA, LG, and FT ranged from 45 to 60 ms at E18 and E20. Regression analyses revealed that SA, LG, FT and TA burst durations did not vary with TA cycle duration. Both slope and R2 were approximately 0.0 in 89% of all burst duration analyses (N = 76 regressions, E18 and E20 combined).

Foot-Free

Following removal of egg shell anterior to the foot (foot-free), E20 embryos extended the leg beyond the egg and produced larger joint excursion ranges during some RLMs (9 embryos). Compare the joint angle amplitudes for 2 RLMs from a single experiment during control (Figure 5A) and foot-free conditions (Figure

5B), noting differences in scale. There was also an increase in cycle frequencies greater than 6 Hz. However, few differences were observed in EMG patterns between conditions. For example, in the experiment shown, SA burst onset latencies mostly fell between 0–0.1 s, weakly varying with cycle duration during both conditions (Figures 5A1, 5B1). Similar SA results were obtained in 6 of 6 foot-free experiments. LG burst onset varied closely with cycle duration during control and foot-free RLMs (Fig. 5A2, 5B2). LG onset closely varied with cycle duration and a late relative onset was the predominant pattern in 4 of 6 foot-free experiments. Participation rates for foot-free experiments (Figure 5C) were also similar to E20 control data (Figure 1D).

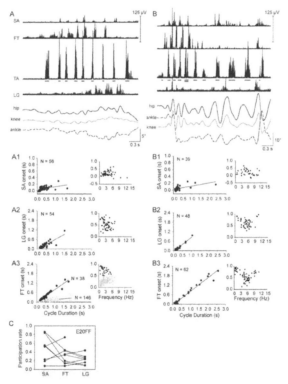

Figure 5. Control and foot-free RLMs during an E20 experiment. A: A control RLM of 8 cycles is shown (4.0±1 Hz). B: A double horizontal line identifies the end of one foot-free RLM (6.8±0.7 Hz) and start of another much slower RLM (3.1±0.7 Hz). Note differences in kinematic scales and also the faster frequencies during foot-free RLMs. A1and B1: SA onset latencies were similar for control (A1: slope = 0.08, R2 = 0.20) and foot-free RLMs (B1: slope = 0.09, R2 = 0.38). Relative onsets for SA were also similar. A2 and B2: LG onset strongly co-varied with cycle duration in both conditions (A2: slope = 0.63, R2 = 0.73; B2: slope = 0.89, R2 = 0.94). A3 and B3: Control FT burst onset (A3) displayed 2 regression trends (open circles: slope = 0.10, R2 = 0.30; closed circles: slope = 0.80, R2 = 0.93) that sorted by relative onset (open circles 0–0.5; closed circles 0.5–1.0). FT onset during foot-free RLM (B3) closely varied with cycle duration (slope = 0.86, R2 = 0.97), and exhibited a bias for late relative onset. C: SA, FT, and LG participation rates are shown for 9 foot-free experiments. The SA implant was lost in one experiment and the LG was in another.

FT onset was more likely to vary closely with TA cycle duration during foot-free experiments (5 of 7) than E20 control experiments (5 of 10). For example, FT burst onsets for the control data in the experiment shown in Figure 5 suggested there were 2 recruitment patterns (open and closed circles) across the entire range in cycle duration (Figure 5A3, left plot). The relative onset plot (Figure 5A3, right plot) revealed that the burst onsets sorted as early (open circles) and late (closed circles) in the TA cycle. The onset latency for early bursts was weakly associated with TA cycle duration; whereas, onset for late bursts was strongly associated. During foot-free RLM, early bursts were no longer observed at lowest RLM frequencies (Figure 5B3, right plot). Collectively, FT exhibited a distinct relative onset bias (>60% of sample) in 9 of 10 control experiments, and a preference for early onset (7 experiments). In contrast, only 4 of 7 foot-free experiments exhibited an FT relative onset bias, 2 early and 2 late. FT onset was also more likely to overlap TA activity during foot-free RLMs (33±11% of cycles) compared to control (15±15%, p<0.011), but burst durations (TA 65 ms, FT 55 ms) and duration of overlap (≈ 20 ms) were similar between conditions.

Interlimb Alternation during RLMs

Given some RLM features appeared to be similar to locomotor patterns while others were more ambiguous, we implanted leg muscles bilaterally at E20 in 4 experiments to determine if both legs were active during RLMs. We found that all experiments exhibited evidence of interlimb coordination. For example, in Figure 6A, the right TA and LG were alternately active, as were the right and left TA, and the right LG alternated with and the left flexor digitorum profundus (toe flexor/ankle extensor, FDP). The plot of ankle markers tracking anterior-posterior displacements indicated that the legs were moving in opposite directions, as during fore-aft foot trajectories while walking. Burst analyses for this experiment indicated that 56% of all cycles for the right LG (N = 211) were accompanied by bursts in the left LG or FDP. Results indicated that 50% of left FDP bursts began mid cycle (0.4–0.6). In the 4 experiments combined, EMG bursts were detected in the left leg during 49% to 62% of rhythmic sequences in the right leg, and the majority of left bursts began mid cycle (Figure 6B). A plot of burst frequencies for each leg (4 experiments) verified that these interlimb events fell within the RLM frequency range (Figure 6C).

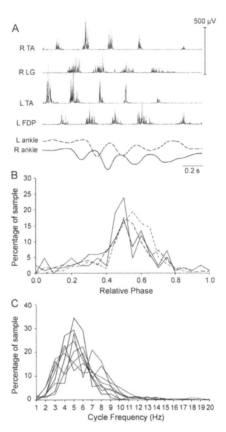

Figure 6. Interlimb coordinated during RLM at E20. Interlimb data are shown for 4 experiments at E20. A: A 3-burst sequence in the right (R) TA (3.4±1.2 Hz) alternated with a 4-burst sequence in the left (L) TA (4.0±0.7 Hz). R TA bursts also alternated with R LG bursts; L TA bursts alternated with left flexor digitorum profundus (L FDP) bursts. Anterior (upward)/posterior (downward) displacements of the 2 ankles were out of phase with one another. B: Left ankle extensor burst onsets were normalized to right extensor burst cycles (relative phase) and phase values were binned in 0.05 increments before plotting the distribution of interlimb phase patterns. The majority of left leg bursts fell between relative phases of 0.4–0.6 in all 4 experiments. C: The burst frequency distributions for the cycles plotted in B are shown for each leg (L TA and R TA). TA burst frequencies ranged from 1 to 10 Hz for both legs in each experiment.

Discussion

RLM Patterns Resemble Locomotor Patterns

Our first aim was to determine if EMG patterns for RLMs resemble any of the three locomotor patterns in hatchings. During walking, swimming and airstepping, several features are characteristic of leg muscle activity [1], [2]. Nearly all leg muscles exclusively participate in either the flexor synergy for limb protraction or extensor synergy for limb retraction. Burst onsets are close in time for flexor synergists (i.e., TA, SA). The ankle extensor (LG) alternates with flexors and the overlap between the bursts is little or none. In addition, activity of the knee extensor (FT) is distinctly different from other leg extensors. FT bursts twice per step cycle in walking and swimming, and once during airstepping; one FT burst begins during flexor activity in all 3 forms of locomotion. Our results indicated that EMG patterns for RLMs shared these features. SA burst onsets were closely associated with TA onset, regardless of TA burst frequency (Figure 3B). The ankle extensor, LG, was consistently recruited in the interval between TA bursts (Figures 1, 3). Recruitment of the knee extensor, FT, was distinct from LG; it most often began early in the cycle, but it could also burst late or twice per cycle (Figures 3–5). Also, like locomotion, RLMs could include alternating interlimb coordination (Figure 6). This array of similarities leads us to propose that RLMs in late-stage embryos are locomotor-related behavior.

Was there a match between the RLM burst pattern and any of the 3 locomotor forms at hatching? The RLM frequency range (1–10 Hz) embraced the combined cycle frequencies for the 3 locomotor behaviors at hatching [2]. On average, RLM frequencies at E18 (4 Hz) and E20 (5.4 Hz) were faster than the approximate averages for walking (2.8 Hz) and swimming (3.3 Hz), but airstepping (5.1 Hz) includes the higher frequencies common to RLMs. Though studies of locomotor patterns for hatchlings used the LG as reference for analyses, it also appears that the relative timing of burst onsets during RLMs closely resembled airstepping EMG [1], [2]. During RLMs, LG began in the latter half of the TA cycle and was immediately followed by the next TA burst (Figures 1, 3). FT began near TA offset and before onset of the next LG burst (Figures 1, 4–5). Extrapolating from published averages for airstepping (see [2] Figure 5), LG appears to begin in the later 40% of the TA cycle. A single FT burst begins near the end of the TA burst and terminates during the LG burst [2]. Further, during airstepping extensor burst durations are brief (<100 ms) and vary independent of cycle duration, a feature that was also typical of RLMs.

Features that Distinguish RLMs from Locomotor Behaviors

RLM EMG patterns also differed from airstepping. RLMs included burst frequencies below 3.5 Hz, a broader spectrum of FT onsets, and short lasting TA bursts that did not co-vary with cycle duration. Whereas during airstepping, TA burst durations are longer and scale with the step cycle [2]. The absence of covariation between extensor burst and cycle duration also distinguished RLMs from locomotion and swimming, for extensor bursts during stance (locomotion) and limb retraction (swimming) closely co-vary with the step cycle [2]. Thus, the absence of scaling between burst and cycle duration distinguished RLMs from all

3 posthatching forms of locomotion. The absence of burst scaling during RLMs could be an indication that proprioception at E18–E20 is not sufficiently mature to code movement dynamics [24]. However, longer extensor bursts are observed during hatching at E20 [17], possibly as embryos extend their legs into the shell wall to rotate their posture [25]. Thus, another possibility is that during most RLMs, the leg was not adequately loaded by foot contact with the shell to modulate extensor activity, for there were a few instances when LG burst durations were modestly lengthened (Figures 5A, 6A). On the other hand, flexor burst durations were probably not modulated by a load because flexor muscles operated in a shortened range due to the postural constraint imposed by the shell. It is also possible that descending pathways are required to modulate burst duration in ovo but that they do not modulate RLM activity.

Shell Removal did not Enhance Locomotor Features of RLMs

We also considered the possibility that the extreme flexed posture imposed by shell constraint could mediate a set of cues that might mask locomotor patterns for airstepping, walking and/or swimming, and match force output to the limited movement space. The extreme flexed posture in ovo during the final days of incubation puts flexors at their most shortened length and all leg extensors at their greatest length. Thus our second aim was to determine if EMG patterns would appear more similar to one or more locomotor patterns when the shell constraint was removed. We reasoned that if leg movements were unconstrained, movements would be larger, plus flexor muscle lengths and loads would increase as extensor lengths decreased. However our results indicated that the temporal structure of RLM EMG activity was preserved, despite substantial increases in joint excursion range after shell removal (Figure 5B) and a significant increase in cycle frequency [23]. The resilience of the pattern during larger excursions was consistent with the observation that TA and SA burst onsets co-varied even when the ankle rotated out of phase with the hip and knee during control RLM (Figures 1B, 3A). FT onset was the only measure that exhibited a modest variation with shell removal. FT onset was more likely to co-vary with TA cycle duration, but relative onset was more ambiguous or shifted from an early to late burst in the TA cycle (Figure 5B3); and the incidence of briefly overlapping TA and FT activity significantly increased.

It was somewhat surprising to us that the EMG pattern for RLMs at E20 was not more mutable during the larger unconstrained motions, given the embryo would hatch and walk within 24 hrs. Any of the possible mechanisms discussed above might account for the minimal effect of shell removal. Also, the sensory cues for larger motions may have been insufficient because the embryo was in a side lying position and the leg flexors experienced less gravitational loading than if held vertically suspended, as during airstepping studies. Further, shell removal did not appreciably alter head/neck/spine flexion, and these inputs might act to reduce descending drive to spinal motor networks during foot-free as well as control experiments. On the other hand, it may be instructive that recruitment of FT, the muscle with the most varied pattern during control RLMs, was also the only muscle responsive to motion-related changes. The early and late onset variability in FT seems to be comparable to the onset shifts also seen in the vastus lateralis, a knee extensor in cat, during the ballistic motions of paw shaking [26], [27].

RLM Pattern and Rhythm have Different Developmental Time Frames

In our previous study we propose that RLM rhythm generation may be modular because stable rhythms appeared to be isolated to individual muscles [23]. The same array and distribution of rhythms were observed by E15. Our present findings extend these results, suggesting that rhythm is established within a motor pool before it is expressed by an array of motor pools to form the RLM pattern. Participation of two muscles was infrequent at E15 but well established at E18. Studies of fictive locomotion and scratching in adult preparations have also observed that rhythmic activity in motor neuron pools can be sustained or resumed without temporal resetting as some motor pools drop out [28]–[31]. These observations led to proposals that central pattern generation for rhythmic limb movements involves two or more levels of control, i.e., one for rhythm and another for pattern (see [32] for a recent review). Thus, the differences in developmental time frame found for RLM rhythm and pattern generation in our studies appear to provide support for the proposal that pattern and rhythm generation involve multiple levels of control.

The developmental time frames for rhythm and pattern generation extend even more broadly than addressed by our present data. Repetitive limb movements are slow (0.2–2 Hz) at E9 [20], and temporally irregular between E12 and E15 [17], [19], [22]. The slow rhythms at E9 are likely produced by transient spinal network dynamics resulting from immature membrane properties [33], [34], and their progressive maturation may partially account for the temporal irregularities preceding emergence of RLM rhythms at E15. We do not know the mechanisms responsible for RLM rhythms, but descending pathways are well established by E15 [35], [36], and might provide excitatory drive as maturing spinal neurons lose their intrinsic excitability. Both slow and fast rhythms are also expressed in adult lamprey locomotion [37]. The slow frequencies are attributed to neural membrane properties, and the fast frequencies are attributed to recurrent excitation within the neural network (see [38] for a recent review). Our results suggest that the RLM pattern appears 1 to 3 days after RLM rhythm, however RLM pattern development may begin with the emergence of alternating flexor and extensor synergies by E9 [18]. Some of the subsequent transformations are likely attributable to maturation of membrane properties such as persistent inward currents [39]. However, the emergence of the FT double burst and its variants at E18 may require development of additional mechanisms such as another control layer in CPG circuitry [32]. Maturation of sensory inputs may also shape FT burst patterns [2]; however FT double bursting is retained in fictive locomotion after hatching, suggesting that it is centrally specified [40].

In sum, our present study of RLM burst patterns completes the first detailed studies of RLMs in the chick embryo. Findings revealed that despite considerable variations in the ensemble of active muscles during RLMs, burst patterns adhered to a common template. The template consisted of alternating flexion and extensor synergies, but also included burst patterns that distinguished knee extensor activity from ankle extensor activity. These findings are consistent with essential features of locomotor behaviors in hatchlings, and appear most similar to airstepping. Thus we conclude that RLMs in the late stage embryo are locomotor-related and produced by the developing neural circuits for locomotion. Our findings also suggest that RLM patterns are only modestly impacted by small modifications in physical constraint. Whether more dramatic changes in environmental conditions and postural context can significantly impact late stage embryonic behavior remains to be explored. Finally, our results lead us to conclude that RLM rhythm and pattern have different developmental time frames and that further study of their development may advance our understanding of locomotor control.

Materials and Methods

Fertile Leghorn chicken eggs were incubated under standard conditions with a temperature-controlled humidified incubator (37.5°). Eggs were maintained under similar conditions in a temperature-controlled chamber during preparation and recording. The first day of incubation was E0 and the present studies were conducted at E15, E18 or E20. Age was verified at the end of experiments using staging criteria by Hamburger and Hamilton [41]. The viability of the embryo was assessed throughout the experiment by visually monitoring pulse rate and tracking body displacements with a force transducer (FT03C, Grass Instruments) placed in contact with the thigh [42]. If either indicator suggested the embryo was deteriorating, the experiment was terminated. All procedures were approved by the University Institutional Animal Care and Use Committee.

EMG, Video and Force Recordings

Embryos were prepared for synchronized electromyographic (EMG) and video recording of leg movements in the egg. A sagittal view of the leg was obtained by creating a window in the shell and dissecting egg membranes. Leg muscles were implanted with silver bipolar electrodes (o.d. 50 μm), including the sartorius (SA), a hip flexor; femorotibialis (FT), a knee extensor; tibialis anterior (TA), an ankle dorsiflexor; and lateral gastrocnemius (LG), an ankle extensor. The location of EMG electrode tips was later verified by dissection after euthanizing the embryo. In E18 and E20 embryos, modified minutin pins were inserted along the leg for digitizing hip, knee and ankle movements. In some experiments additional shell was removed anterior to the foot to reduce mechanical constraints on leg posture and movement (foot-free).

Synchronized EMG, force and video were recorded continuously for ≥3–4 hours (Datapac 2K2, Run Technologies) on a grounded anti-vibration table. EMG was band pass filtered (100–1,000 Hz), amplified (×2000), and digitally sampled (4 kHz). Force signals were low pass filtered (30 Hz), amplified (×20,000) and digitized (500 Hz).

RLM Analyses

Sequences of repetitive muscle bursting that were accompanied by leg movement were selected for analyses. The synchronized video and/or force recordings were used to confirm the presence of leg movements. The force recording was particularly useful for detecting RLM when joint motions were only a few degrees in magnitude due to spatial constraint but of sufficient force to produce rhythmic displacements of the body. EMG signals were rectified to detect burst onsets and offsets based on 3 criteria: burst threshold (2–4 times baseline amplitude), burst duration (20–400 ms), and inter-burst interrupt (20 ms) (Datapac 2K2). Baseline amplitude for each EMG channel was estimated from a quiescent 100 ms segment of recording at the start of the experiment. The threshold for each EMG channel was determined by pilot analyses of several RLMs and based upon the combined parameter set that captured the greatest number of EMG bursts. An example of the burst detection method is shown in Figure 7.

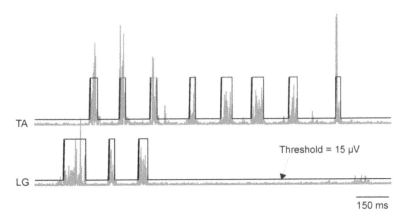

Figure 7. Burst detection methods. After signal rectification and baseline offset adjustment to 0 μV, average baseline channel noise was estimated during a quiescent interval of 100 ms. In the above example, a threshold of 15 μV (3 times baseline) for TA and LG burst detection is shown. EMG activity remaining above threshold for 20–400 ms and surrounded by subthreshold signal for 20 ms or longer was defined as a burst. Examples of burst detection are indicated by the upward square wave defections in the threshold trace.

EMG burst and cycle measures were referenced to TA burst onset, because TA was the most reliably rhythmic of the muscles recorded [23]. Thus, TA cycle duration was defined as the time between consecutive TA burst onsets. Burst onset latencies for all other muscles were measured from onset of the preceding TA burst, with the exception of SA bursts that slightly preceded the concurrent TA burst. Latencies were divided by the concurrent TA cycle duration to obtain relative onset latencies. The time between offset of TA bursts and onset of antagonist muscles (FT, LG) was also calculated to determine if alternating bursts exhibited some duration of co-activity with TA. Leg movement during the RLM was digitized and joint angles were calculated for display with the EMG traces using methods established for in ovo recordings [22], [43].

Analyses were limited to RLM cycles that were rhythmically stable (SD≤1 Hz) for 4 or more consecutive TA bursts [23]. Regression analyses were used to examine linear trends in EMG burst parameters relative to TA cycle duration within an experiment, and coefficients of determination (R2) are reported. Data sets containing fewer than 20 bursts were excluded. The t-test for independent samples was used to compare subject means between age groups (E18 and E20) and conditions (control and foot-free). A p<0.05 was considered significant.

Acknowledgements

We wish to express our appreciation to Shon Carney and Hilary Genise for their assistance in data analyses. We are very grateful for helpful comments on an earlier draft of this paper by Drs. Doug Stuart, Nicolas Schweighofer and Linda Fetters, as well as significant encouragement for this work offered by Dr. Robert Gregor.

Authors' Contributions

Conceived and designed the experiments: NSB. Performed the experiments: NSB. Analyzed the data: YUR NSB. Contributed reagents/materials/analysis tools: YUR. Wrote the paper: YUR NSB.

References

1. Jacobson RD, Hollyday M (1982) A behavioral and electromyographic study of walking in the chick. J Neurophysiol 48: 238–256.

2. Johnston RM, Bekoff A (1996) Patterns of muscle activity during different behaviors in chicks: implications for neural control. J Comp Physiol [A] 179: 169–184.

3. Muir GD, Gosline JM, Steeves JD (1996) Ontogeny of bipedal locomotion: Walking and running in the chick. J Physiol (Lond) 493: 589–601.

4. Hamburger V (1963) Some aspects of the embryology of behavior. Q Rev Biol 38: 342–365.

5. Hamburger V, Balaban M, Oppenheim R, Wenger E (1965) Periodic motility of normal and spinal chick embryos between 8 and 17 days of incubation. J Exp Zool 159: 1–14.

6. Hamburger V, Wenger E, Oppenheim R (1966) Motility in the chick embryo in the absence of sensory input. J Exp Zool 162: 133–160.

7. Bekoff A, Kauer JA (1984) Neural control of hatching: fate of the pattern generator for the leg movements of hatching in posthatching chicks. J Neurosci 4: 2659–2666.

8. Bekoff A, Sabichi AL (1987) Sensory control of the initiation of hatching in chicks: effects of a local anesthetic injected into the neck. Dev Psychobiol 20: 489–495.

9. O'Donovan M, Chub N (1997) Population behavior and self-organization in the genesis of spontaneous rhythmic activity by developing spinal networks. Semin Cell Dev Biol 8: 21–28.

10. Tabak J, Senn W, O'Donovan M, Rinzel J (2000) Modeling of spontaneous activity in developing spinal cord using activity-dependent depression in an excitatory network. J Neurosci 20: 3041–3056.

11. Milner LD, Landmesser LT (1999) Cholinergic and GABAergic inputs drive patterned spontaneous motoneuron activity before target contact. J Neurosci 19: 3007–3022.

12. Hanson MG, Landmesser LT (2004) Normal patterns of spontaneous activity are required for correct motor axon guidance and the expression of specific guidance molecules. Neuron 43: 687–701.

13. Hanson MG, Landmesser LT (2006) Increasing the frequency of spontaneous rhythmic activity disrupts pool-specific axon fasciculation and pathfinding of embryonic spinal motoneurons. J Neurosci 26: 12769–12780.

14. Myers CP, Lewcock JW, Hanson MG, Gosgnach S, Aimone JB, et al. (2005) Cholinergic input is required during embryonic development to mediate proper assembly of spinal locomotor circuits. Neuron 46: 37–49.

15. Haverkamp LJ (1986) Anatomical and physiological development of the xenopus embryonic motor system in the absence of neural activity. J Neurosci 6: 1338–1348.

16. Haverkamp LJ, Oppenheim RW (1986) Behavioral development in the absence of neural activity: effects of chronic immobilization on amphibian embryos. J Neurosci 6: 1332–1337.

17. Bekoff A (1976) Ontogeny of leg motor output in the chick embryo: a neural analysis. Brain Res 106: 271–291.

18. Bradley NS, Bekoff A (1990) Development of coordinated movement in chicks. 1. Temporal analysis of hindlimb muscle synergies at embryonic days 9 and 10. Dev Psychobiol 23: 763–782.

19. Bradley N (1999) Transformations in embryonic motility in chick: kinematic correlates of type I and II motility at E9 and E12. J Neurophysiol 81: 1486–1494.

20. Chambers SH, Bradley NS, Orosz MD (1995) Kinematic analysis of wing and leg movements for type I motility in E9 chick embryos. Exp Brain Res 103: 218–226.

21. Sharp AA, Ma E, Bekoff A (1999) Developmental changes in leg coordination of the chick at embryonic days 9, 11, and 13: uncoupling of ankle movements. J Neurophysiol 82: 2406–2414.

22. Bradley NS, Solanki D, Zhao D (2005) Limb movements during embryonic development in the chick: evidence for a continuum in limb motor control antecedent to locomotion. J Neurophysiol 94: 4401–4411.

23. Bradley NS, Ryu YU, Lin J (2008) Fast locomotor burst generation in late stage embryonic motility. J Neurophysiol 99: 1733–1742.

24. Maier A (1993) Development of chicken intrafusal muscle fibers. Cell Tissue Research 274: 383–391.

25. Hamburger V, Oppenheim RE (1967) Prehatching motility and hatching behavior in the chick. J Exp Zool 166: 171–204.

26. Koshland GF, Smith JL (1989) Mutable and Immutable Features of Paw-Shake Responses After Hindlimb Deafferentation in the Cat. J Neurophysiol 62: 162–173.

27. Hoy MG, Zernicke RF, Smith JL (1985) Contrasting roles of inertial and muscle moments at the knee and ankle during paw-shake response. J Neurophysiol 54: 1282–1294.

28. Lafreniere-Roula M, McCrea D (2005) Deletions of rhythmic motoneuron activity during fictive locomotion and scratch provide clues to the organization of the Mammalian central pattern generator. J Neurophysiol 94: 1120–1132.

29. Stein P (2005) Neuronal control of turtle hindlimb motor rhythms. J Comp Physiol A Neuroethol Sens Neural Behav Physiol 191: 213–229.

30. Stein PS, Daniels-McQueen S (2002) Modular organization of turtle spinal interneurons during normal and deletion fictive rostral scratching. J Neurosci 22: 6800–6809.

31. Stein PS, Daniels-McQueen S (2004) Variations in motor patterns during fictive rostral scratching in the turtle: knee-related deletions. J Neurophysiol 91: 2380–2384.

32. McCrea DA, Rybak IA (2008) Organization of mammalian locomotor rhythm and pattern generation. Brain Res Rev 57: 134–146.

33. Chub N, O'Donovan M (1998) Blockade and recovery of spontaneous rhythmic activity after application of neurotransmitter antagonists to spinal networks of the chick embryo. J Neurosci 18: 294–306.

34. Chub N, Mentis GZ, O'Donovan MJ (2006) Chloride-sensitive MEQ fluorescence in chick embryo motoneurons following manipulations of chloride and during spontaneous network activity. J Neurophysiol 95: 323–330.

35. Okado N, Oppenheim RW (1985) The onset and development of descending pathways to the spinal cord in the chick embryo. J Comp Neurol 232: 143–161.

36. Glover JC (1993) The Development of Brain Stem Projections to the Spinal Cord in the Chicken Embryo. Brain Res Bull 30: 265–271.

37. Cangiano L, Grillner S (2003) Fast and slow locomotor burst generation in the hemispinal cord of the lamprey. J Neurophysiol 89: 2931–2942.

38. Grillner S (2006) Biological pattern generation: the cellular and computational logic of networks in motion. Neuron 52: 751–766.

39. Cotel F, Antri M, Barthe JY, Orsal D (2009) Identified ankle extensor and flexor motoneurons display different firing profiles in the neonatal rat. J Neurosci 29: 2748–2753.

40. Jacobson RD, Hollyday M (1982) Electrically evoked walking and fictive locomotion in the chick. J Neurophysiol 48: 257–270.

41. Hamburger V, Hamilton H (1992) A Series of Normal Stages in the Development of the Chick Embryo (Reprinted from Journal of Morphology, Vol 88, 1951). Dev Dyn 195: 231–272.

42. Bradley NS, Jahng DY (2003) Selective Effects of Light Exposure on Distribution of Motility in the Chick Embryo at E18. J Neurophysiol.

43. Orosz MD, Bradley NS, Chambers SH (1994) Correcting two-dimensional kinematic errors for chick embryonic movements in ovo. Comput Biol Med 24: 305–314.

CITATION

Ryu YU and Bradley NS. Precocious Locomotor Behavior Begins in the Egg: Development of Leg Muscle Patterns for Stepping in the Chick. PLoS ONE 4(7): e6111. doi:10.1371/journal.pone.0006111. Copyright © 2009 Ryu, Bradley. Originally published under the Creative Commons Attribution License, http://creativecommons.org/licenses/by/3.0/

Perinatal Androgens and Adult Behavior Vary with Nestling Social System in Siblicidal Boobies

Martina S. Müller, Julius F. Brennecke, Elaine T. Porter,
Mary Ann Ottinger and David J. Anderson

ABSTRACT

Background

Exposure to androgens early in development, while activating adaptive aggressive behavior, may also exert long-lasting effects on non-target components of phenotype. Here we compare these organizational effects of perinatal androgens in closely related Nazca (Sula granti) and blue-footed (S. nebouxii) boobies that differ in neonatal social system. The older of two Nazca booby hatchlings unconditionally attacks and ejects the younger from the nest within days of hatching, while blue-footed booby neonates lack lethal aggression. Both Nazca booby chicks facultatively upregulate testosterone (T) during fights, motivating the prediction that baseline androgen levels differ between obligately siblicidal and other species.

Methodology/Principal Findings

We show that obligately siblicidal Nazca boobies hatch with higher circulating androgen levels than do facultatively siblicidal blue-footed boobies, providing comparative evidence of the role of androgens in sociality. Although androgens confer a short-term benefit of increased aggression to Nazca booby neonates, exposure to elevated androgen levels during this sensitive period in development can also induce long-term organizational effects on behavior or morphology. Adult Nazca boobies show evidence of organizational effects of early androgen exposure in aberrant adult behavior: they visit unattended non-familial chicks in the colony and direct mixtures of aggression, affiliative, and sexual behavior toward them. In a longitudinal analysis, we found that the most active Non-parental Adult Visitors (NAVs) were those with a history of siblicidal behavior as a neonate, suggesting that the tendency to show social interest in chicks is programd, in part, by the high perinatal androgens associated with obligate siblicide. Data from closely related blue-footed boobies provide comparative support for this interpretation. Lacking obligate siblicide, they hatch with a corresponding low androgen level, and blue-footed booby adults show a much lower frequency of NAV behavior and a lower probability of behaving aggressively during NAV interactions. This species difference in adult social behavior appears to have roots in both pleiotropic and experiential effects of nestling social system.

Conclusions/Significance

Our results indicate that Nazca boobies experience life-long consequences of androgenic preparation for an early battle to the death.

Introduction

The diversity of vertebrate social systems has motivated many studies regarding their evolutionary origins, but few have identified the key proximate controls. The central role of the endocrine system is well recognized, functioning at the interface of the social environment and the physiology of organisms. It orchestrates behavioral responses to social cues in real time and the development of socially relevant behavioral and morphological phenotypes in ontogenic time [1]. The available data from behavioral endocrinology suggest the possibility that early exposure to androgens is required to trigger the attacks characteristic of the young of some vertebrates [2]. Studies of birds in particular have recently supported the position that nestling social interactions are mediated by circulating androgens of maternal or endogenous origin [3]. The data from developmental endocrinology suggest that such exposure, if it occurs during developmentally sensitive periods, could induce organizational effects [4] on behavior [5], [6] and/or morphology [7]–[9] that represent pleiotropic epiphenomena. To our knowledge, these two areas of study involving hormones have not been integrated to ask the question: when androgens are mobilized in the context of sibling aggression in developing vertebrates, do those androgens also induce enduring phenotypes affecting later life stages?

Here we explore the endocrine dynamics behind two divergent neonatal social systems in closely related seabird species [10], testing predictions regarding the regulation of androgens just after hatching, when lethal aggression is expressed in one of the species, and the potential long-term consequences of this exposure. Nazca boobies (Sula granti) raise only one offspring at a time, as do other pelagic seabirds, but often lay a second egg that counters the poor hatching success characteristic of this species [11], [12]. Obligate siblicide solves the frequent problem of hatching both eggs: the older nestling (A-chick) unconditionally attacks its younger, smaller sibling (B-chick) and ejects it from the nest within days of hatching [13]. In facultatively siblicidal blue-footed boobies (Sula nebouxii), fatal sibling aggression is conditional on food availability, and, if it occurs, does so later in the nestling period [14], [15]. Nazca booby neonates thus face a high probability of lethal fighting, while blue-footed booby neonates do not.

Obligate siblicide represents the maximum challenge posed in an aggressive contest: two neonates are confined together until one kills the other. The dramatic fitness consequences of poor performance in this social system should lead to strong selection for adaptations related to siblicide in Nazca boobies, such as an endocrine milieu that facilitates the rapid onset of aggressive behavior. During fights, both Nazca booby hatchlings facultatively upregulate testosterone [T; 3], in accord with the Challenge Hypothesis [2], implicating testosterone in a quick and forceful transition to combat. While potentially critical to facilitating a rapid mobilization of T for aggression, high perinatal androgens can also induce long-term phenotypic effects, such as impaired immune function, compromised future reproduction, and developmental instabilities [8], [16]–[18]. Following the demonstration that both first- and second-hatching Nazca booby nestlings temporarily up-regulate T during fights [3], we predicted that Nazca booby neonates also maintain a higher baseline level of potentially costly androgens than do species lacking obligate siblicide. Blue-footed boobies provide the complement for a powerful comparative test of this idea, given their phylogenetic [10], morphological, behavioral, and ecological [19] similarities to Nazca boobies, including siblicidal behavior [13]. Blue-footed booby neonates receive no benefits of androgen-based aggression at hatching, so the long-term costs of early exposure to these agents should cause selection to penalize high circulating androgens related to neonatal aggression. We compare hatchling androgen levels in the two species to ask whether Nazca boobies hatch with higher concentrations of androgens than do blue-footed boobies.

Elevated steroid concentrations during developmentally sensitive periods can have organizational effects on the central nervous system and program certain facets of behavior; the effect of uterine androgen exposure on adult social phenotype of mice is a well known example [20]. Unusual social interest of non-breeding adult Nazca boobies in conspecific and heterospecific nestlings [21], [22] provides an opportunity to examine organizing effects of androgens on behavior. These Non-parental Adult Visitors (NAVs) search the breeding colony for unguarded nestlings, join them at the nest, and display parental/courtship behavior, aggression, and sexual behavior in various mixtures [21]. No satisfactory ultimate explanation for this phenomenon exists; NAV behavior confers no obvious fitness benefits on individuals [23], yet is ubiquitous among both sexes at our study site [21] and elsewhere in the species' range [Malpelo Island, Colombia, F. Estela, pers. comm]. On the proximate level, NAVs were reported to have higher corticosterone levels and lower T than non-NAVs, consistent with an activational role in this behavior [24]. We used complete life histories of birds in our long-term study population in a longitudinal test of the hypothesis that the neonatal T surges that accompany siblicide-related aggression induce organizational effects on the tendency to show NAV behavior in adulthood, by asking whether siblicidal individuals perform NAV behaviors at a higher frequency than do nonsiblicidal individuals.

While attendance of active nests by nonbreeders is a well-documented aspect of prospecting behavior in many bird species [25], this intense social interest of adults in unrelated young is virtually unreported elsewhere in the literature [21], although it may occur in other species at low frequency, until now misidentified as poor parental behavior or as anomalous. We present the first report of NAV behavior in the closely related blue-footed booby (Sula nebouxii) from observations performed in a dense breeding colony on Isla Lobos de Tierra, Perú. Our previous understanding of blue-footed booby behavior comes from a breeding colony in the Galápagos Islands, with inter-nest distances ranging between 3.3—377 m [26]. NAV behavior was observed on only one occasion during 46 person-years of field work in and around that colony. In Perú, however, blue-footed booby colony density was comparable to the density of the Nazca booby nests where we studied them in the colony at Punta Cevallos, Española Island, Galápagos [27], [28]. Ecological factors, such as high nesting density, may cause some populations to show NAV behavior while others do not. Here, we compare NAV behavior of the two species from similar colony environments, integrating these data into the model of behavioral organization via early androgen exposure. With the contrast in neonatal androgen level established (see Results), we compare relative frequency of NAV behavior among Nazca boobies and blue-footed boobies and examine different types of NAV behaviors separately to test the following hypothesis: if androgens associated with siblicide organize NAV aggression, then blue-footed boobies should show NAV behavior less frequently and less aggressively than do Nazca boobies.

In summary, we use both longitudinal analysis and the comparative method to examine linkage among neonatal social system, androgen concentrations, and enduring behavioral effects of early exposure to androgens. We test two hypotheses: that Nazca boobies hatch with higher baseline androgen levels than do blue-footed boobies, in parallel with the contrast in lethal fighting among hatchlings; and that participation in the different neonatal social systems induces persistent developmental consequences for behavior.

Results

Androgens

All nestlings providing androgen (combined 5α-DHT and T) samples came from one- or two-egg clutches, and were categorized into four types: product of single-egg clutch, product of a two-egg clutch in which only one egg hatched, first nestling in a two-nestling brood, and second nestling in a two-nestling brood. ANOVA of androgen level, incorporating species, nestling type, and sex effects, and their interactions, revealed a significant species effect (Table 1, Fig. 1). Nestling history and some interaction terms also explained significant variation in androgen level (Table 1), but the species difference was the largest component of variance by far, with an ANOVA mean square exceeding that of the second largest component (nestling history) by a factor of 14 (Table 1). Removal of the non-significant effects from the model simplified the result, rendering the species * nestling history * sex interaction non-significant ($F_{3,131} = 2.147$, P = 0.10), and maintaining the strong species effect ($F_{,131} = 78.814$, P<10^{-6}) and lesser nestling history ($F_{3,131} = 6.702$, P = 0.0003) and nestling history * sex effects ($F_{1,131} = 2.954$, P = 0.035). With the exception of second-hatching nestlings, which typically are killed shortly after hatching, Nazca booby neonates had three times or more the androgen level of comparable blue-footed boobies (Fig. 1).

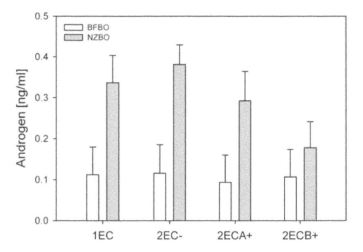

Figure 1. Mean (±95% CI) androgen level in Nazca booby (NZBO) and blue-footed booby (BFBO) hatchlings. 1EC = products of one-egg clutches, 2EC- = products of two-egg clutch where only one egg hatched, 2ECA+ = first hatchling from two-egg clutch where both eggs hatch, 2ECB+ = second hatchling from two-egg clutches where both eggs hatch.

Table 1. ANOVA results comparing androgen levels of Nazca and blue-footed boobies, incorporating effects of nestling type and sex and their interactions.

Effect	d.f	MS	F	p
Intercept	1	488.106	1412.126	$<10^{-6}$
Species	1	27.884	80.672	$<10^{-6}$
Nestling history	3	1.925	5.570	0.001
Sex	1	1.198	3.466	0.065
Species * Nestling history	3	0.748	2.163	0.096
Species * Sex	1	0.441	1.275	0.261
Nestling history * Sex	3	1.289	3.728	0.013
Species*Nestling history * Sex	3	0.981	2.839	0.041
Error	126	0.346		

Behavior

Longitudinal data from Nazca boobies derived from two-egg clutches showed that siblicidal nestlings mature into adults that exhibit more NAV behavior than do non-siblicidal conspecifics (t-value = 2.468, df = 104, N = 108, P = 0.015; Fig. 2A). Specifically, siblicidal Nazca boobies displayed more frequent aggressive NAV behavior as adults (t-value = 2.287, df = 104, N = 108, P = 0.024; Fig. 2B), whereas the frequencies of affiliative and sexual NAV behavior did not vary significantly with nestling social experience (t-value = 1.304, df = 104, N = 108, P = 0.195; t-value = 0.006, df = 104, N = 108, P = 0.995).

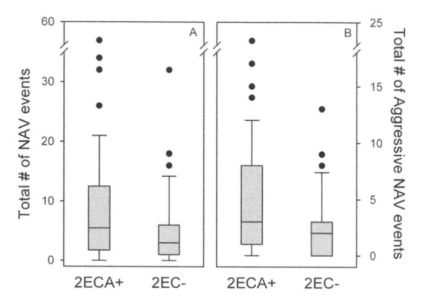

Figure 2. Number of NAV events (A: total, B: aggressive) performed by adults derived from two-egg clutches. Siblicidal individuals (2ECA+) are contrasted with non-siblicidal individuals (2EC-) whose sibling egg failed to hatch. Boxes and whiskers represent the middle 50% and middle 80% of the data, respectively.

We detected a far lower frequency of NAV behavior in Peruvian blue-footed boobies. When expressed as the number of NAV events per non-breeder present, scaled to observation effort, Nazca and blue-footed boobies performed 0.0725 (95% CI = 0.0588–0.0862) and 0.0017 (95% CI = 0.0014–0.0018) NAV events/ nonbreeder/hour, respectively. Scaled to the number of unattended chicks available, we observed 0.0586 (95% CI = 0.0.0489–0.0683) and 0.0152 (95% CI = 0.0138–0.0166) NAV events/unattended chick/hour in Nazca boobies and blue-footed boobies, respectively. In each of these comparisons, neither mean is included in the other mean's 95% CI. We predicted that if early androgen exposure organizes aggressive NAV behavior in boobies, then aggressive NAV interactions should represent a smaller fraction of the NAV repertoire in blue-footed boobies than in Nazca boobies. We found support for this hypothesis: in most years, Nazca booby NAV events included significantly more aggressive and less affiliative behaviors compared to blue-footed booby events (Table 2; log-linear analysis, Nazca booby frequencies differ from those of blue-footed boobies [$P << \alpha_{crit}$] in all years except 2001–02 [P = 0.24]).

Table 2. Percentage of behavior types observed in NAV events, by species.

Nazca boobies

Year	Affiliative	Aggressive	Sexual	N
2000–01	34.4	49.5	16.0	2169
2001–02	45.0	41.9	13.1	1263
2002–03	34.1	57.5	8.4	2069
2004–05	37.1	56.7	6.3	2638

Blue-footed boobies

Year	Affiliative	Aggressive	Sexual	N
2006–07	50.3	35.2	14.5	165

Most Nazca booby chicks show physical evidence of past aggressive NAV attacks, including lacerations and bare patches of skin where down has been scraped away [21]. In contrast, we found 0 out of 1473 medium-sized blue-footed booby chicks had abrasions or missing down, suggesting that while aggressive NAV interactions occur in blue-footed boobies, they are a milder form of the phenomenon. The absence of such markers is apparently not attributable to higher durability of blue-footed boobies' plumage or skin, because Nazca booby NAVs visit blue-footed booby chicks at our Galápagos study site and cause visible injuries to them [22].

Discussion

We found that hatchling Nazca boobies, facing a strong possibility of an imminent fight to the death, emerge from the egg with a higher circulating androgen level than do blue-footed boobies, providing comparative evidence of the role of androgens in sociality. With the exception of B-chicks, which typically are killed shortly after hatching, Nazca booby neonates had three times or more the androgen level of comparable blue-footed boobies. Among these Nazca booby hatchlings, androgen level was not sensitive to the post-natal stimuli of the presence of a potential competitor (an egg) or to position in the laying sequence (Fig. 1), suggesting that levels on the day of hatching reflect a species difference in exposure to androgens extending back in time into the embryonic period, rather than facultative upregulation after hatching.

Within the Nazca booby population, more than half of the clutches contain two eggs in most years [29], [30], and 68% of two-egg clutches produce two hatchlings [11], so more than 1/3 of all Nazca boobies experience the additional early androgen exposure associated with upregulation of T during obligate siblicide

[3]. Given the evidence that neonatal birds with altricial development experience organizational effects from post-hatching exposure (the first 30% of the nestling period) to steroid hormones [31], the early exposure of highly altricial Nazca boobies to a high androgen level is expected to coincide with a sensitive period in development. Accordingly, we found evidence of organizational effects of siblicide-related androgens in aberrant adult behavior. Siblicidal Nazca boobies showed more frequent social attraction (NAV behavior) to unguarded chicks years later as a nonbreeding adult, compared to non-siblicidal Nazca boobies (Fig. 2A). This predictive ability of siblicidal history was due to a significant association with the frequency of aggressive NAV interactions (Fig. 2B); it did not predict the frequencies of the affiliative or sexual components of NAV behavior. These results provide rare evidence from a natural population that social behavior (in this case, aggressive interest in non-familial young) expressed as an adult is conditioned by androgen exposure while young, which itself is related to the extreme social challenge faced by neonatal Nazca boobies.

Nazca boobies provide a rare example of a non-human animal known to express such widespread and varied social interest in unrelated young. Adolescent male, but not female, yellow-rumped caciques (Cacicus cela) persistently attack and attempt to copulate with fledglings, with unknown fitness consequences [32]. Black-legged kittiwake (Rissa tridactyla), nonbreeders also visit active nests where parents are absent, although this "squatting" behavior is interpreted as an assessment of site quality, important for territory acquisition [33]. Neither of these cases nor more anecdotal reports seem to represent parallels to the Nazca booby situation, in which the motivation to visit active nests is clearly social interaction with a chick, with fully mature adults of both sexes engaging in a variety of social behaviors. The aggressive component of NAV behavior of Nazca boobies shows an ontogenic linkage with androgen exposure experienced during siblicide events. This result prompts the expectation that the elevated androgens of Nazca boobies before siblicide, compared to blue-footed boobies (Fig. 1) and probably most other birds, also lead to the later expression of NAV behavior. Consistent with this expectation, NAV behavior occurs at low frequency in Peruvian (this study) and Galápagos (personal observation) blue-footed booby populations, and is unreported from central American colonies. Moreover, aggression represents a lower proportion of the NAV repertoire of blue-footed boobies compared to Nazca boobies (Table 1). The contrast in the NAV phenomenon among the two species is consistent with the hypothesis that perinatal androgens organize propensity to show NAV behavior, in a behavioral cascade rooted in the hatchling social environment. Experimental manipulation of perinatal androgen exposure in these species, with longitudinal followup of adult social behavior, can evaluate this idea further.

Launching their infants into vastly different social systems, Nazca and blue-footed boobies otherwise exhibit similar ecologies, life histories, and phylogenies, minimizing potentially confounding variables that would interfere with identification of the proximate regulators of these social systems. These species also provide an opportunity to assess life-long consequences of neonatal social system dynamics on phenotype via these proximate mechanisms. Here we demonstrate links between neonatal social system, androgen concentrations, and persistent phenotypic effects of early exposure to androgens. We suggest that the strong selection for the androgens that facilitate obligately siblicidal behavior in Nazca boobies outweighs any pleiotropic consequences arising later in life.

Materials and Methods

Blood Sampling and Assays

As part of a long-term study on this species we monitored over 16,000 Nazca booby nests at Punta Cevallos, Isla Española, (89° 37' W, 1° 23' S) in the Galápagos Islands since 1984 in which laying dates, laying order, clutch size, hatching dates, and hatching order were recorded. We sampled first (CA) and second (CB) chicks from 15 nests with two egg clutches where both eggs hatched (2ECA+ and 2ECB+, respectively), 15 nests with two egg clutches where either the first egg or the second egg was present at hatching of the other egg but did not hatch itself (2EC-), and 15 chicks from one-egg clutches (1EC). Chicks were sampled within 24 hours of hatching, usually between 10am–1pm. Approximately 200 µl of blood was taken from the brachial vein using a 27 ½ gauge needle and unheparinized microcapillary tubes and then collected in 1.5 mL Eppendorf microcentrifuge tubes. Typically, sampling time was 3 min., in rare cases 5–8 min. Samples were processed within 1–2.5 hrs of collection. We centrifuged blood samples in the field for 10 minutes, then removed a known volume of serum which was stored separately in tubes containing 750 µl absolute EtOH. Red blood cells were resuspended in ca. 500 µl of 70% EtOH. All samples were stored at ambient temperature for 1 month before refrigeration upon arrival at Wake Forest University. Hormones were extracted and assayed at the University of Maryland. DNA was extracted from red blood cells for PCR sex identification at Wake Forest University [34].

We monitored 925 blue-footed booby nests in Dec 2006 on Isla Lobos de Tierra (80° 51' W, 6° 24' S), Perú which already contained eggs, using only nests that contained one- or two-egg clutches. We collected blood samples from chicks within 24 hours of hatching using the same procedure used for Nazca boobies. We sampled 16 first and second hatchlings from two-egg clutches (2ECA+ and

2ECB+, respectively), 16 chicks from two-egg clutches where one egg did not hatch (2EC-), and 16 chicks from one-egg clutches (1EC). Hormone samples were extracted at Wake Forest University and assayed at the University of Maryland, DNA was extracted from red blood cells for PCR at Wake Forest University.

After double ether steroid extraction (85% recovery), androgens (5α-DHT and T) were assayed via radioimmunassay (RIA; 35) for chicks from both species. The RIA was validated for parallelism, sensitivity (10pg/ml), accuracy, and precision (<10% CV) for serum from both species.

Behavioral Observations

Nazca Boobies

In January 2005, we placed distinctive blue plastic bands on all adult nonbreeders and failed breeders that had known nestling histories (age, clutch size, hatching order and siblicidal/non-siblicidal) in the "study area," a subsection of the Nazca booby colony at Punta Cevallos [28]. The numbers on the bands could be read easily from a distance. From January 19, 2005–Mar 31, 2005, each afternoon between ca. 1300–1630 hrs, two observers systematically patrolled the study area recording all NAV events, the identity of the NAVs, and the nest number of the chick victims (504 person/hrs in total). Any NAV interaction of the three classes between a plastic-banded nonbreeder and a chick was recorded as a NAV event; repeated NAV interactions of a given behavior type between the same individuals in one day were all considered as the same "NAV event" [23]. NAV behavior classes included the following: aggressive (biting, shaking, or jabbing), affiliative (attending the chick with little interaction, preening, presenting gifts of feathers or pebbles), or sexual (attempted copulation with chick). Some analyses used an extended data set to include NAV observations recorded during the breeding seasons of 2001–02 and 2002–03 using the same methods.

In 2004–05, an average of 33.0 banded Nazca booby nonbreeders (SD = 17.0) was present at mid-day in the observation area over the course of the NAV observation period. During the subsequent afternoon hours, we observed an average of 9.22 active NAVs (SD = 5.1) in the study area and a mean of 1.85 NAV events per person-hour (SD = 1.28; n = 1750 NAV events). The study area had an average of 130.3 medium-sized chicks in the colony (>than 20 days old; <fledging age), 66.8 of which (SD = 19.2) were unattended on average.

Correct assessment of the relative frequency of NAV behavior required adjustment for variation in colony attendance among individuals. We performed nightly band re-sight surveys of the study area in which we noted all plastic banded birds.

Only birds present on at least 5 nights during the period of the NAV observations were considered colony residents and all others were excluded from the analysis. In addition, a bird had to be of "nonbreeder" status on at least one of the nights present to be included in the analysis as a potential NAV.

Blue-Footed Boobies

We conducted systematic behavioral observations in a large blue-footed booby colony on Isla Lobos de Tierra, Perú (80° 51' W, 6° 24' S) in Dec 2006. Patrolling a subsection of the colony, we recorded all NAV behaviors and categorized them as aggressive, affiliative, or sexual, and noted the time of day. We performed 30 hours of behavioral observations over the course of five days (29 Dec 2006 to 2 Jan 2007; Day 1: 1000–1600 hrs; Days 2–4: 930–1630 hrs; Day 5: 930–1230 hrs). We patrolled an area with hundreds of unattended medium-sized chicks (between 20–100 days old), and even higher numbers of non-breeding adults, in which abundant opportunity for NAV interactions existed. Because adults were unbanded and so unrecognizable, we were unable to determine how many, if any, adults were involved in repeated NAV events with a particular chick. As a consequence, our assumption that each NAV event was independent may not be strictly true. To perform a single loop of the area required 30 minutes; for each pass of the colony we recorded each NAV interaction with a given chick as a new event due to the ample number of nonbreeders in the colony and relative high probability that the NAV was a new bird. We distinguished affiliative NAVs from parents by observing the chicks' responses to the attending adults, as in Nazca booby chicks [21]: blue-footed booby chick victims of NAV behavior tuck their bills under and push their forehead to the ground, leaving the backs of their necks exposed in a submissive posture, during the interactions.

An average of 1950.7 nonbreeders (SD = 90.1) were present during the middle of the day in the observation area during the five days of behavioral observations, and we observed a mean of 3.23 NAV events per person-hour (SD = 0.33, n = 165 NAV events). Chick attendance by blue-footed booby parents (85.6%) was higher than by Nazca boobies (44.4%), which may have limited opportunity for blue-footed booby nonbreeders to show NAV behavior. However medium-sized chicks (>20 days, <fledging age) were abundant: the observation area had 1473 medium-sized chicks and an average of 212.3 (SD = 31) chicks that were unattended, or three times the number of available Nazca boobies at Punta Cevallos.

Statistical Analyses

We compared androgen levels using a three-way ANOVA (main effects species, sex, and nestling type, after checking homogeneity of variances using Levene's

Test ($F15$, $126 = 1.73$, $P>0.05$) and normality using a normal probability plot of residuals. Untransformed data showed a tendency to non-normality, which was corrected by log-transformation, and we used log-transformed androgen level in the ANOVA.

To test the organizational effects hypothesis with data from Nazca boobies, using R [38], we fitted a generalized linear mixed model to frequency of total NAV events of different types performed by an individual, using nestling history, sex, and age as fixed effects and number of nights presents in the colony as a nonbreeder as a random effect. Nestling history was a dichotomous variable describing whether or not a nonbreeder derived from a two-egg clutch was siblicidal as a chick. The residuals of the dependent variables followed a negative binomial, or "zero-inflated Poisson" distribution, so we specified the model for an "overdispersed Poisson" distribution. Using the t-value output, we determined significance with a two-tailed test with $\alpha = 0.05$.

To compare the frequencies of NAV behavior types across species, we first used a log-linear analysis on Nazca booby data alone to test for temporal variation across years, and found a significant year x behavior type interaction (Maximum Likelihood $\chi2 = 202.05$, $df = 6$, $P<0.01$). As a result, we did not collapse all Nazca booby data into a single sample for comparison with the single year of data from blue-footed boobies, instead comparing each year of Nazca booby data with the single year of blue-footed booby, adjusting the P values for multiple comparisons with the false discovery method [36], [37]. The false discovery method computes a critical α level (αcrit) for each comparison, to which the P value is compared to determine significance.

Acknowledgements

We thank the Galápagos National Park Service for permission to work in the Park; the Charles Darwin Research Station, TAME Airline, Carlos Zavalaga, and Ecoventura/Doris Welch for logistical support; our many assistants and colleagues, especially A. Gunderson, K. Birchler, and J. Awkerman, for assistance in producing our long-term databases; K. Whitehouse for lab assistance; and T. Maness, V. Apanius, K. Huyvaert, and E. Lavoie for statistical and editorial advice.

Authors' Contributions

Conceived and designed the experiments: DA MM EP. Performed the experiments: JB DA MM. Analyzed the data: DA MM MO. Contributed reagents/materials/analysis tools: DA MO. Wrote the paper: DA MM.

References

1. Pfaff D, Arnold A, Etgen A, Fahrbach S, Rubin R, editors. (2002) Hormones, Brain and Behavior. Elsevier Academic Press.

2. Wingfield JC, Hegner RE, Duffy AM, Ball GF (1990) The challenge-hypothesis: theoretical implications for patterns of testosterone secretion, mating systems, and breeding strategies. Amer Nat 136: 829–846.

3. Ferree ED, Wikelski MC, Anderson DJ (2004) Hormonal correlates of siblicide in Nazca boobies: support for the Challenge Hypothesis. Horm Behav 46: 655–662.

4. Phoenix CH, Goy RW, Gerall AA, Young WC (1959) Organizing action of prenatally administered testosterone propionate on the tissues mediating mating behavior in the female guinea pig. Endocrinology 65: 369–382.

5. Adkins EK (1975) Hormonal basis of sexual differentiation in the Japanese quail. J Comp Physiol Psychol 89: 61–71.

6. Adkins EK (1979) Effect of embryonic treatment with estradiol or testosterone on sexual differentiation of the quail brain: critical period and dose-response relationships. Neuroendocrinology 29: 178–185.

7. Strasser R, Schwabl H (2004) Yolk testosterone organizes behavior and male plumage coloration in house sparrows (Passer domesticus). Behav Ecol Sociobiol 56: 491–497.

8. Rubolini D, Romano M, Martinelli R, Leoni B, Saino N (2006) Effects of prenatal yolk androgens on armaments and ornaments of the ring-necked pheasant. Behav Ecol Sociobiol 59: 549–560.

9. Romano M, Rubolini D, Martinelli R, Alquati AB, Saino N (2005) Experimental manipulation of yolk testosterone affects digit length ratios in the ring-necked pheasant (Phasianus colchicus). Horm Behav 48: 342–346.

10. Friesen VL, Anderson DJ (1997) Phylogeny and evolution of Sulidae (Aves: Pelecaniformes): A test of alternative modes of speciation. Mol Phylogenet Evol 7: 252–260.

11. Humphries CD, Arevalo VD, Fischer KN, Anderson DJ (2006) Contributions of marginal offspring to reproductive success of Nazca booby (Sula granti) parents; tests of multiple hypotheses. Oecologia 147: 379–390.

12. Anderson DJ (1990) Evolution of obligate siblicide in boobies: A test of the Insurance-Egg Hypothesis. Amer Nat 135: 334–350.

13. Anderson DJ (1989) The role of hatching asynchrony in siblicidal brood reduction of two booby species. Behav Ecol Sociobiol 25: 363–368.

14. Drummond H, Gonzalez E, Osorno JL (1986) Parent-offspring cooperation in the blue-footed booby (Sula nebouxii) - social roles in infanticidal brood reduction. Behav Ecol Sociob 19: 365–372.

15. Lougheed LW, Anderson DJ (1999) Parent blue-footed boobies suppress siblicidal behavior of offspring. Behav Ecol Sociob 45: 11–18.

16. Navara KJ, Hill GE, Mendonca MT (2005) Variable effects of yolk androgens on growth, survival, and immunity in eastern bluebird nestlings. Physiol Biochem Zool 78: 570–578.

17. Rubolini D, Martinelli R, von Engelhardt N, Romano M, Groothuis TGG, et al. (2007) Consequences of prenatal androgen exposure for the reproductive performance of female pheasants (Phasianus colchicus). Proc R Soc B 274: 137–142.

18. Uller T, Ekloef J, Anderson S (2005) Female egg investment in relation to male sexual traits and the potential for transgenerational effects in sexual selection. Behav Ecol Sociobiol 57: 584–590.

19. Nelson JB (1978) The Sulidae: Gannets and Boobies. Oxford, UK: Oxford University Press.

20. Vom Saal FS (1989) Sexual differentiation in litter-bearing mammals: Influence of sex of adjacent fetuses in utero. J Anim Sci 67: 1824–1840.

21. Anderson DJ, Porter ET, Ferree ED (2004) Non-breeding Nazca boobies (Sula granti) show social and sexual interest in chicks: behavioral and ecological aspects. Behavior 141: 959–977.

22. Townsend HM, Huyvaert KP, Hodum PJ, Anderson DJ (2002) Nesting distributions of Galápagos boobies (Aves: Sulidae): an apparent case of amensalism. Oecologia 132: 419–427.

23. Porter ET (2003) Non-breeding Nazca boobies (Sula granti) show social and sexual interest in chicks: causes and consequences. Unpublished master's thesis, Wake Forest University, Winston-Salem, NC.

24. Tarlow EM, Wikelski M, Anderson DJ (2003) Correlation between plasma steroids and chick visits by non-breeding adult Nazca boobies. Horm Behav 43: 402–407.

25. Danchin E, Cadiou B, Monnat JY, Rodriguez Estrella R (1991) Recruitment in long-lived birds : conceptual framework and behavioral mechanisms. Proc Int Ornithol Congr 20: 1641–1656.

26. Anderson DJ, Hodum PJ (1993) Predator behavior favors clumped nesting in an oceanic seabird. Ecology 74: 2462–2464.

27. Anderson DJ (1993) Masked Booby (Sula dactylatra). In: Poole A, Gill F, editors. Philadelphia: Washington, D. C.: The Academy of Natural Sciences. The American Ornithologists' Union. The Birds of North America, No. 73.

28. Apanius V, Westbrock MA, Anderson DJ (2008) Reproduction and immune homeostasis in a long-lived seabird, the Nazca booby (Sula granti). Ornithol Monogr 65:

29. Anderson DJ (1990) Evolution of obligate siblicide in boobies. 1: A test of the insurance egg hypothesis. Am Nat 135: 334–350.

30. Clifford LD, Anderson DJ (2001) Food limitation explains most clutch size variation in the Nazca booby. J Anim Ecol 70: 539–545.

31. Adkins-Regan E, Mansukhani V, Seiwert C, Thompson R (1994) Sexual differentiation of brain and behavior in the zebra finch: critical periods for effects of early estrogen treatment. Journal of Neurobiology 25: 865–877.

32. Robinson SK (1988) Anti-social behavior of adolescent yellow-rumped caciques (Icterinae: Cacicus cela). Anim Behav 36: 1482–1495.

33. Cadiou B, Monnat JY, Danchin E (1994) Prospecting in the kittiwake, Rissa tridactyla: different behavioral patterns and the role of squatting in recruitment. Anim Behav 47: 847–856.

34. Maness TJ, Westbrock MA, Anderson DJ (2007) Ontogenic sex ratio variation in Nazca boobies ends in male-biased adult sex ratio. Waterbirds 30: 10–16.

35. Ottinger MA, Pitts S, Abdelnabi MA (2001) Steroid hormones during embryonic development in Japanese quail: Plasma, gonadal, and adrenal levels. Poul Sci 80: 795–799.

36. Curran-Everett D (2000) Multiple comparisons: philosophies and illustrations. American Journal of Physiology Regulatory Integrative and Comparative Physiology 279: R1–R8.

37. Benjamini Y, Hochberg Y (1995) Controlling the false discovery rate – a practical and powerful approach to multiple testing. Journal of the Royal Statistical Society B – Methodological 57: 289–300.

38. Faraway JJ (2006) Extending the Linear Model with R: Generalized Linear Mixed Effects and Nonparametric Regression Models. Boca Raton, FL: Chapman & Hall/CRC.

CITATION

Müller MS, Brennecke JF, Porter ET, Ottinger MA, and Anderson DJ. Perinatal Androgens and Adult Behavior Vary with Nestling Social System in Siblicidal Boobies. PLoS ONE 3(6): e2460. doi:10.1371/journal.pone.0002460.

Transcriptomic Profiling of Central Nervous System Regions in Three Species of Honey Bee During Dance Communication Behavior

Moushumi Sen Sarma, Sandra L. Rodriguez-Zas, Feng Hong, Sheng Zhong and Gene E. Robinson

ABSTRACT

Background

We conducted a large-scale transcriptomic profiling of selected regions of the central nervous system (CNS) across three species of honey bees, in foragers that were performing dance behavior to communicate to their nestmates the location, direction and profitability of an attractive floral resource. We used microarrays to measure gene expression in bees from Apis mellifera, dorsata and florea, species that share major traits unique to the genus and also show striking differences in biology and dance communication. The goals of this study were to determine the extent of regional specialization in gene expression and to explore the molecular basis of dance communication.

Principal Findings

This "snapshot" of the honey bee CNS during dance behavior provides strong evidence for both species-consistent and species-specific differences in gene expression. Gene expression profiles in the mushroom bodies consistently showed the biggest differences relative to the other CNS regions. There were strong similarities in gene expression between the central brain and the second thoracic ganglion across all three species; many of the genes were related to metabolism and energy production. We also obtained gene expression differences between CNS regions that varied by species: A. mellifera differed the most, while dorsata and florea tended to be more similar.

Significance

Species differences in gene expression perhaps mirror known differences in nesting habit, ecology and dance behavior between mellifera, florea and dorsata. Species-specific differences in gene expression in selected CNS regions that relate to synaptic activity and motor control provide particularly attractive candidate genes to explain the differences in dance behavior exhibited by these three honey bee species. Similarities between central brain and thoracic ganglion provide a unique perspective on the potential coupling of these two motor-related regions during dance behavior and perhaps provide a snapshot of the energy intensive process of dance output generation. Mushroom body results reflect known roles for this region in the regulation of learning, memory and rhythmic behavior.

Introduction

Animal brains are composed of anatomically distinct regions which are further made up of spatially and functionally coherent populations of neurons and glia. They specialize in processing different kinds of signal input from the animal's internal and external environment and integrate the information to mount an appropriate physiological and behavioral response. Even though many molecular processes are considered universal to all cells, transcriptomics and in situ hybridization analysis have revealed extensive localized regulation of genes expressed in the brain in both vertebrates and invertebrates [1]–[3]. Studies of mammals and song birds have revealed strong connections between brain-region specific gene expression and behavior [4], [5].

The brain of the honey bee, Apis mellifera, is among the best studied insect brains, from neuroanatomical, neurochemical and neurophysiological perspectives [6]–[8]. In addition, numerous brain-region specific analyses of gene expression exist for the honey bee, but they are largely limited to analyses of single genes via in situ analysis [9]–[11]. Although honey bees have been used for several large-scale analyses of behaviorally related gene expression at the whole brain level [12]–[14], large-scale transcriptomic comparisons of different brain regions in the bee brain have not yet been conducted. This information would be helpful to our understanding of how known regional differences in structure and function in the bee brain relate to behavioral regulation.

We performed the current study with two goals in mind. Firstly, to carry out a transcriptomic profiling of selected regions of the honey bee brain to determine the extent of regional specialization in gene expression. A recent neuroanatomical analysis [15] of dance language [16], the famous communication system used by honey bee foragers to communicate to their nestmates the location, direction and profitability of an attractive food source they encounter in the environment, suggested that multiple brain regions are involved in the perception and production of dance communication, meaning that regional analysis of brain gene expression will be required to understand this remarkable system. Therefore, our second goal was to explore the honey bee CNS at the transcription level to get a picture of how the different regions might contribute to the behavioral output associated with dance communication.

Honey bee foragers need to carry out a spectrum of sensory information processing not only to navigate but also to produce the dance language. These include visual information about the landscape and location, direction information, measurement of distance, measurement of gravity to name a few. Based on previous neuroanatomical and behavioral studies in honey bees and other insects, we know that the following CNS regions are likely to be involved in sensory processing and regulation of dance: 1) the optic lobes (OL), which receive sensory input from the compound eyes and the ocelli and are comprised of 3 distinct neuropils, the lamina, medulla and lobula [17]–[19]; 2) the mushroom bodies (MB), which consist of intrinsic neurons called Kenyon cells [20], [21] and a complex neuropil arranged into anatomically defined subparts strongly associated with olfactory learning, higher order visual processing, multi-modal sensory integration and general arousal [22]–[29]; and 3) the central brain (CB), which contains (among other neuropils) the central complex [30], a precisely arranged array of neurons implicated in the control of acoustic communication and coordinated movements during courtship in Drosophila (fruit fly) and gomphocerine grasshoppers [31]–[33], orientation to polarized light [34], [35]. We also included the second thoracic ganglion (TG) because it innervates and controls the body parts involved in the dance output namely, the wings, the middle and hind legs, muscles of meso and metathorax and the articulation of the abdomen with the thorax through the propodeum [36]. The TG has also been implicated in coordinating motor patterns, generating rhythmic movements in flies and crickets and gregarious behavior in locusts [37]–[39].

We exploited the striking differences in dance language that exist in the genus Apis [16], focusing on three species, A. mellifera, A. dorsata, and A. florea. A. mellifera, the cavity nesting Western honey bee, the model honey bee species for which we have the genome sequence and related genomic resources [40], is the species in which the dance language was first described. The other two species that are confined mostly to South Asia show some striking differences in the dance language [16]. Our previous study showed differences in gene expression between these species [13], but the study was conducted on whole brains, and more importantly, it compared foragers and one-day-old bees, so it was not clear to what extent the differences were related to differences in dance behavior or differences in behavioral maturation.

We generated CNS region-specific profiles of gene expression for A. mellifera, dorsata, and florea individuals sampled directly from beehives while they were engaged in dance behavior. We were particularly interested in testing for two types of patterns of CNS regional gene expression in association with dance behavior. Differences in gene expression between brain regions that are consistent across the three bee species should reflect intrinsic functional specialization within the Apis nervous system. By contrast, regional differences that are different across the three bee species (region by species interactions) may reflect differences that are related to species differences in behavior.

Methods

Sample Collection and Processing

Dancing bees returning from successful pollen collecting trips were easily identified on honeycombs according to established criteria [41] and collected from 2–4 natural colonies on location in Bangalore, India between 9 AM and 12 PM each collection day. Individuals were collected on liquid nitrogen and subsequently stored in ultra-low freezers. Samples were shipped on dry ice to the University of Illinois and stored at –80°C until processed further. 2 colonies from each species were used for subsequent analysis. Frozen brains were fixed in RNALater ICE (Ambion/Applied Biosystems, Austin, Texas) and dissections were carried out on fresh ice under a stereomicroscope (Olympus SXZ12). Fig. 1 shows the meridians along which the brain was divided to give the 3 brain regions studied. Due to limitations of the technique the divisions were not precise and might have missed cell bodies that lie at the junction of two regions, e.g. some cell bodies that lie close to the antennal lobes and send their projections into the central complex might have been removed along with the antennal lobes [30]. However, a majority of the cells that belong to a particular region were included. In order to include the central complex in the central brain region, we could only have the calyces of the mushroom bodies in the MB region. However, the calyces contain the cell bodies of the intrinsic Kenyon cells [20] where most (but not all) transcription takes place.

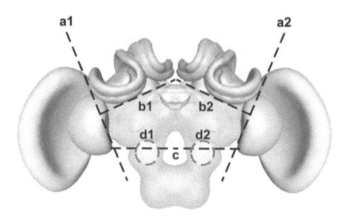

Figure 1. Schematic representation of brain with the regions that were used in the study. Dotted lines show the meridians of separation between the regions: a1 and a2–optic lobes, b1 and b2–mushroom bodies, c, d1 and d2 mark the lines along which the sub-esophageal ganglion and the antennal lobes were removed. Brain schema in Figure 1, 2, 3 and 5 drawn after [71].

Extractions were carried out with RNAeasy (Qiagen, Valencia, California) kit and quantified using a Nanodrop™ spectrophotometer (Thermo Scientific, Wilmington, Delaware). 100 ng of each RNA sample was amplified using the MessageAmp kit (Ambion/Applied Biosystems, Austin, Texas). Amplified mRNA from OL, MB, CB and TG of each individual dancer (11–12 individuals/colony/species) were used in labeling and hybridization as in previous studies [13].

Microarray Analysis

We analyzed 4 CNS regions of 72 pollen dancers of 3 species on an A. mellifera brain EST microarray. This array has been shown to perform well for these species even though it was designed with mellifera sequences [13]. A loop design was employed for microarray analysis [42], with each CNS region compared to another region belonging to the same species, on multiple arrays per species. A total of 117 arrays were used in this study, each probing equal quantities of amplified mRNA (2 ug). CNS regions from individual bees were hybridized on each array.

Data Analysis

Microarray data generated in this study meet Minimum Information about Microarray Experiment (MIAME) standards and are available at ArrayExpress [43] under accession number E-TABM-700. A total of 117 arrays were used for statistical analysis, after quality control analysis. Microarray features that received a "-100" flag by the scanning software GenePix or that had a median fluorescence intensity <300 were removed from the analysis [44]. Gene expression measurements were log2-transformed and normalized using a LOWESS smoothing function. Microarray elements with missing information in more than two arrays or control sequences [44] were removed from the analysis. Data from duplicated spots were averaged and adjusted for global dye and microarray effects [45], [46]. In order to minimize errors and the occurrence of false positives, only genes that were expressed at detectable levels in at least 115 arrays of the quality tested 117 microarrays were included in the data analysis. Thus 5182 or 74% of the genes on the arrays that passed the filter criteria can be considered to be ubiquitously expressed throughout the honey bee CNS, irrespective of species. The dataset for each species was then analyzed in two ways, separately subject to ANOVA (ANOVA 1) and combined in a single dataset before being subject to an ANOVA (ANOVA 2).

A linear mixed effect ANOVA model was used to describe the normalized expression intensity (y_{jklmn} or y_{ijklmn}) on a gene-basis: ANOVA 1: y_{jklmn} = $\mu+R_j+D_k+A_l+B_m+H_n+\varepsilon_{jklmn}$; ANOVA 2: y_{ijklmn} = $\mu+S_i+R_j+SR_{ij}+D_k+A_l+B_m+H_n+\varepsilon_{ijklmn}$ where μ denotes the overall mean, S_i denotes the effect of the ith species, R_j denotes the effect of the jth region, D_k denotes the kth dye, A_l denotes the effect of the lth array, B_m denotes the effect of the mth array batch, Hn denotes the effect of the nth bee, and ε_{jklmn} or ε_{ijklmn} denotes the residual. The terms H_n, A_l and e_{jklmn} or e_{ijklmn} were treated as random effects and the remaining terms were treated as fixed effects. Statistical tests were based on a global variance model (F3). The false discovery rate criterion was used to adjust for multiple testing [47]. Statistical analyses were conducted using the SAS statistical package.

Results of a subsequent post-hoc t test were then used to carry out the subsequent pattern analysis. Using a cut-off p value of 10^{-4} we coded a negative expression ratio (log2 fold change) between any two regions as –1, while a positive expression ratio was coded as 1. A non-significant expression difference was coded as 0. Expression profiles that compared all six possible contrasts MB-CB, CB-OL,

CB-TG, MB-OL, MB-TG and OL-TG were then used to cluster genes using a K-means clustering program. Contrasts that compared MB with another region gave the best clustering outcome and therefore only those 3 contrasts CB-MB, OL-MB and TG- MB were used for subsequent pattern analysis. 27 possible patterns of gene expression profiles are possible in these 3 contrasts as summarized in Table 1. Depending on the expression profiles that the genes had in each species, they were grouped into one of the 27 patterns. GO enrichment analysis of genes showing a pattern of interest was carried out with a Chi-square test with Yates continuity correction [12]. Since this correction results in a conservative estimate of the p value, we used a cut-off of $p = 0.01$ for statistical significance. At this threshold, the number of false positives expected was several times lower than the actual significant results obtained. For example, out of 4590 comparisons that were carried out for genes that were upregulated in any one CNS region compared to another (irrespective of species), 262 GO terms were identified at the significance level of $p = 0.01$, which is more than 5 times of the expected number of false positives (45.90) .

Table 1. The number of genes that showed each of 27 possible expression patterns.

Pattern#	Pattern			Count of genes in species			In both species			In all 3 species
	CB_MB	OL_MB	TG_MB	AM	AD	AF	AM=AD	AM=AF	AD=AF	AM=AD=AF
1	−1	−1	−1	450	513	525	319	333	349	276
2	−1	−1	0	42	56	25	17	8	7	5
3	−1	−1	1	1	0	0	0	0	0	0
4	−1	0	−1	337	334	165	148	87	73	53
5	−1	0	0	52	63	76	10	4	4	2
6	−1	0	1	0	1	0	0	0	0	0
7	−1	1	−1	1	4	0	1	0	0	0
8	−1	1	0	0	3	0	0	0	0	0
9	−1	1	1	0	0	0	0	0	0	0
10	0	−1	−1	57	75	62	21	9	16	7
11	0	−1	0	136	171	97	51	28	28	16
12	0	−1	1	13	5	0	0	0	0	0
13	0	0	−1	253	192	115	60	22	16	4
14	0	0	0	2177	1889	2894	1392	1767	1566	1205
15	0	0	1	327	292	100	93	51	44	22
16	0	1	−1	11	22	0	2	0	0	0
17	0	1	0	228	330	165	109	70	87	51
18	0	1	1	94	102	85	22	22	18	9
19	1	−1	−1	0	0	0	0	0	0	0
20	1	−1	0	2	5	0	0	0	0	0
21	1	−1	1	8	15	0	3	0	0	0
22	1	0	−1	1	1	1	0	1	0	0
23	1	0	0	80	80	82	11	10	9	2
24	1	0	1	482	482	244	233	118	96	64
25	1	1	−1	1	1	0	1	0	0	0
26	1	1	0	62	62	42	18	9	12	7
27	1	1	1	367	367	416	181	208	206	132

−1 denotes gene expression is higher in MB compared to the other region being compared, 0 denotes equal expression levels, while 1 denotes lower expression level in MB compared to the other region being compared. Abbreviations: CB = central brain, MB = mushroom bodies, OL = optic lobe, TG = thoracic ganglion; AM = A. mellifera, AD = A. dorsata, AF = A. florea.

Results

CNS-Specific Differences in Honey Bee Gene Expression Consistent Across the Species

A total of 5182 genes representing 74% of the genes present on the array passed through our analysis filters (see Methods). About half the genes showed no CNS-specific pattern of expression, presumably reflecting genes involved in processes common to all nervous tissue, across all three species. There were significant differences in gene expression between CNS regions for ca 50% of the genes (ANOVA 1, FDR<0.001; 2597 in mellifera, 2777 in dorsata and 2028 in florea). Approximately 50% of these have been annotated, largely on the basis of known functions in Drosophila melanogaster [40]. The MB was most different from the other CNS regions in gene expression and was thus a major contributor to this region effect. The average proportion of genes differentially expressed in MB was 72% compared to CB (1837 in mellifera, 1949 in dorsata and 1580 in florea), 60% compared to OL (1482 in mellifera, 1663 in dorsata and 1333 in florea) and 82% compared to TG (2177 in mellifera, 2204 in dorsata and 1704 in florea = 1704;). By contrast, the smallest difference in gene expression was observed between CB and TG. The average proportion of genes differentially expressed in TG compared to CB was 14% (461 in mellifera, 422 in dorsata, and 225 in florea).

Similar results were obtained in an independent clustering-based analysis that generated 27 distinct patterns of expression differences between the different CNS regions (Table 1). The biggest gene cluster group (pattern #14) was comprised of genes that showed no region-specific pattern of expression. These genes again presumably reflect genes involved in processes common to all nervous tissue, across all three species. As in the analysis above, ca. 50% of genes showed this pattern in each species (1205 genes). More genes in this category were shared between mellifera and florea than either did with dorsata. Patterns 1 and 27 were the next major groups, wherein MB had a higher or lower expression level respectively compared to the other regions. Again 50% of genes with these patterns were shared between the three species. Genes expressed at similar levels in MB compared to OL but differentially expressed compared to CB and TG were part of the next two major patterns (nos. 4 and 24).

To gain further insight into the possible functional significance of the consistent differences in gene expression between CNS regions across the three species, we performed GO enrichment analyses on the groups of (GO annotated) genes that showed a directional bias of expression in one region compared to another in all 3 species. As with the previous analyses, results for MB compared to CB and TG yielded the most coherent patterns, while comparisons with OL or comparisons between OL, CB and TG did not show concordance between species. Figs. 2

and 3 summarize the results of the enrichment analysis of genes that were upregulated in MB compared to CB and TG, respectively. An almost identical list of GO terms appeared in both comparisons, reflecting consistent themes for MB across the three species. Many of the enriched GO categories pertain to neuronal activity while other categories include those involved in cell surface receptor-linked signal transduction and intracellular signaling cascades, and genes that bind to other proteins (GO molecular function: protein binding).

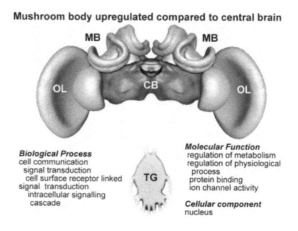

Figure 2. Results of GO enrichment analysis of genes that showed consistent differences in gene expression across the three honey bee species in the mushroom bodies compared to central brain.

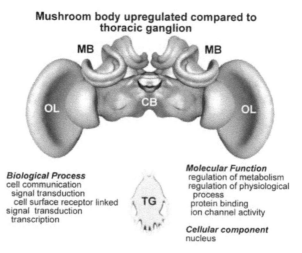

Figure 3. Results of GO enrichment analysis of genes that showed consistent differences in gene expression across the three honey bee species in the mushroom bodies compared to thoracic ganglion.

The following are among the genes upregulated in MB compared to CB and TG in all three species that are known (primarily from functional analysis in Drosophila) to be involved in synaptic transmission: Inositol tris-phosphate receptor (Itp-r83a), known to be preferentially expressed in mushroom bodies of honey bees by in situ hybridization analysis [48], Ryanodine receptor (Drosophila ortholog Rya 44F), Nicotinic acetylcholine receptor and Muscarinic acetylcholine receptor; and Cacophony, a calcium channel gene whose protein product is involved in synaptic transmission that is implicated in Drosophila courtship behavior and adult locomotion, particularly adult male courtship song [49]. The following are among the genes upregulated in MB compared to CB and TG in all three species that are known to be involved in signal transduction: Shaggy, CAMKII known to be highly expressed in honey bee mushroom bodies by in situ hybridization analysis [50], Pka-R2 and Pka-c code for the regulatory and catalytic subunits of cAMP dependent protein kinase or PKA. Shaggy codes for a crucial protein kinase in Drosophila and is an important developmental gene that is also involved in regulation of circadian rhythms in the adult [51]. PKA plays an important role in development and is also involved in adult learning and memory [52], [53]. It is expressed at higher levels in the honey bee mushroom bodies compared to the rest of the brain [54]. Calcium/calmodulin-dependent protein kinase II or CAMKII is involved in learning and memory, specifically long term memory and courtship behavior [55].

In contrast to these results for the MB, we did not detect any concordance in enriched GO categories for genes that are upregulated in OL compared to CB or TG across species. Furthermore, comparatively fewer genes (51 out of 502) in these comparisons showed similar patterns across the species.

CNS-Specific Differences in Honey Bee Gene Expression that Vary by Species

There were significant CNS region by species interactions in gene expression for ca 14% (709 of 5182) of the genes (ANOVA 2, FDR<0.001; Table 2). These reflect regional differences in gene expression that are different across the three bee species. These genes were then subject to a GO enrichment analysis (see Methods). Fig. 4 summarizes cases where there were differences between species in the GO classes that were enriched in genes upregulated in one CNS region compared to another (ANOVA 1). Consistent with the lack of across-species concordance for MB-OL comparisons, there were numerous cases of species-specific MB-OL differences. For example, genes upregulated in OL compared to MB in dorsata were greatly enriched for a number of GO classes that denote involvement in intracellular and cell-cell signaling and regulation of metabolism. On the other hand, mellifera only showed an enrichment of mitochondrial genes upregulated in OL compared to MB while florea by contrast, showed an enrichment of signal transduction genes upregulated in MB compared to OL.

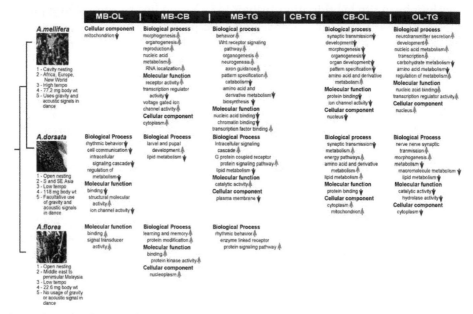

Figure 4. Results of GO enrichment analysis of genes that showed species by CNS region differences in gene expression, based on pair-wise comparisons of the CNS regions (p<0.01, Chi-Square test with Yates continuity correction). First column shows the species with relevant differences in behavior and ecology with phylogenetic ranking after [68], [69]. Upward arrows indicate upregulation of enriched genes of a given GO class in the first brain region of the pair, while downward arrows indicate upregulation of enriched genes of a given GO class in the second brain region of the pair. Abbreviations as in Table 1.

Table 2. Genes showing species by CNS region interaction at p<0.001.

Species\Region	CB	MB	OL	TG
AD_AF	524	516	506	520
AD_AM	541	534	515	523
AF_AM	556	554	542	575

Genes were compared using a post-hoc t-test for differences in expression profiles (p<0.05) for a given CNS region between 2 species. Numbers that showed significant differences are summarized below. Abbreviations as in Table 1.

Fig. 5 summarizes the bias in GO enrichment of genes that were differentially expressed in a given region of one species compared to another species (ANOVA 2). The most biased enrichment was observed primarily for comparisons of mellifera CNS regions with corresponding regions in florea and dorsata. There were many more GO categories for enriched genes upregulated in florea and dorsata CNS regions compared to corresponding regions in mellifera, (32 out of 36 and 26 out of 28 categories enriched in genes differentially expressed in florea and dorsata respectively, compared to mellifera). This is in contrast to the 5 GO categories enriched in florea and dorsata comparisons.

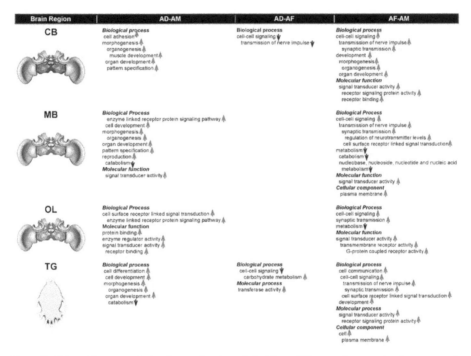

Figure 5. GO enrichment analysis of genes that showed significant differences in expression between species for a given CNS region highlighted in the brain schematic (p<0.04, Chi-Square test with Yates continuity correction). Upward arrows indicate an overrepresentation of upregulated genes of a given GO class in the first species of the pair, while downward arrows indicate overrepresentation of upregulated genes of a given GO class in the second species of the pair. Abbreviations as in Table 1.

Discussion

This "snapshot" of the honey bee CNS during dance behavior revealed some insights into how behavioral differences between species might be reflected in gene expression. The first insight that we gained was that the mushroom bodies were very different from the other CNS regions studied and consistently showed the biggest differences in terms of gene expression. In all three species, the mushroom bodies were the most different from the other regions in terms of gene expression.

In addition, genes involved in signaling and synaptic remodeling were seen to be upregulated compared to other CNS regions. Results from GO analyses highlight the function of the mushroom bodies in learning and memory, with enrichment in categories such as transcriptional regulation and ion channel activity, among others. These results are consistent with known roles for the mushroom bodies in the regulation of rhythmic behavior, learning and memory [28], [29], [33], [56], [57]. In addition, our results nicely correspond with earlier in situ hybridization data and immuno-staining data for genes like Inositol tris-phosphate receptor, CAMKII and PKA that were also shown to be highly expressed in mushroom bodies [48], [50], [54].

Our molecular data provide suggestive evidence for the mushroom bodies being an integration or "association" area in the honey bee CNS [58]. Since we have sampled bees while they were dancing, we are perhaps looking at that part of the CNS that plays the biggest role in processing sensory information and coordinating the dance output. It has been already shown that the small-type Kenyon cells of the mushroom bodies show prominent neural activity in foraging and dancing honey bees [57]. However, 2 alternate possible explanations must also be considered. Firstly, the mushroom bodies are the largest pair of neuropils in the honey bee brain containing 35% of neurons in the honey bee brain. They integrate information from various sensory modalities and thus play a central role in the insect brain [6]. Although we have controlled for the discrepancy in cell numbers between the different regions (see Methods), it is possible that the transcription pattern obtained in mushroom bodies reflects the multimodality of neurons and sensory processing in this part of the CNS. In other words, we are perhaps looking at a chronic difference between mushroom bodies and other parts of the CNS in the honey bee that has nothing to do with the behavior that was being executed at the time of sampling. A third possibility is that the expression profile of honey bee mushroom bodies might be diagnostic of insects in general that have structurally complex mushroom bodies like the hymenopterans (ants, bees and wasps), dictyopterans (cockroaches) and coleopterans (scarab beetles) [59]. Although not closely related, these insects share a marked flexibility in food acquisition behaviors.

Unfortunately, studies on other insects are insufficient for adopting or rejecting any of the 3 scenarios detailed above. There are only two other transcriptomic profiling studies of insect CNS regions and both were carried out on insects that have simpler and smaller mushroom bodies compared to honey bees, Drosophila melanogaster (fruit fly) [60] and Schistocerca gregaria (locust) [61]. Additionally, the animals in those studies were reared in the laboratory and not sampled while carrying out a specific behavior unlike our focal animals. Our approach, applied to other species, might be very useful in exploring the functional significance of region-specific expression in the brain and relating it to evolutionary constraints.

There were far fewer instances of common gene regulation in the optic lobes across the three species. Evidence in other insects links body size to visual ability [62], [63], so the visual systems of the three honey bee species we studied could also be different due to marked differences in size [64]. Of the three species, only dorsata has the ability to fly in very low light conditions. Perhaps reflecting this special ability, the optic lobes showed enrichment of upregulated genes involved in intracellular and cell-cell signaling and regulation of metabolism.

We did not compare dancers with bees carrying out other behaviors because a previous study in honey bees showed that behaviors that are not temporally or physiologically well separated are also not well separated by gene expression [65]. As foragers are very different from workers that stay in the nest [12] the most logical comparison would have been foragers that dance with foragers that do not dance. However this distinction is often ephemeral and not chronic and perhaps more appropriate for quantitative proteomics [66]. Nevertheless, our study provides some hints into the neural and molecular workings of dance behavior. The similarities in gene expression between the central brain and thoracic ganglion provide a unique perspective on the coupling of these two regions during dance behavior. The central brain receives multisensory input like the mushroom bodies does and also coordinates locomotion and rhythmic movement, while the thoracic ganglion receives motor signals from the central brain and provides motor output to the wings, legs and abdomen while generating complex movement patterns [37], [38], [67].

GO analysis reveals that, genes upregulated in both the central brain and thoracic ganglion were similar, mostly dealing with metabolism and energy production. It is likely that these findings reflect the energy intensive process of motor signal transmission and neuronal firing that would be required in generating dance output. If this speculation is correct, then at least some parts of our "snapshot" reflect brain activity that is actually related to dance behavior, rather than to behavior that is regulated over a longer time scale, such as other aspects of foraging behavior. If so, it is worth noting that the two species that showed the most differences in gene expression in the central brain and thoracic ganglion, mellifera and florea, are also the two species that show the biggest differences in dance "dialects," i.e., the precise relationship between dance movements and the distance to the food resource that they encode. This speculation suggests that the central brain and thoracic ganglion gene lists may be particularly valuable for providing candidate genes for distance-related aspects of dance communication.

Apis florea and dorsata showed more CNS-region-specific similarities in gene expression when compared to each other, and both showed more differences when compared to mellifera. This cannot be attributed to evolutionary distance since recent phylogenetic analyses suggest that all three species are separated by 8–10 million years [68], [69]. Instead, we speculate that this reflects the similarities in nesting habit, ecology and dance behavior that exist between florea and dorsata, and not mellifera (Fig. 4) [64]. Both florea and dorsata are open nesting bees that build a single honeycomb from a support, in contrast to mellifera, which is cavity nesting and builds multiple parallel honeycombs inside a tree cavity. Both florea and dorsata are endemic to South Asia and found in primarily tropical and subtropical ecosystems while mellifera is a Western honey bee that is found in both temperate and tropical environments. The dance language also shows striking differences, with florea dancers communicating reportedly exclusively in the visual modality while dorsata is able to use both visual and acoustic signals in its dance communication facultatively. mellifera on the other hand, constrained by the darkness of its hive, communicates with acoustic and vibrational signals. Genes that were upregulated in florea and dorsata central brain compared to mellifera were enriched in GO classes morphogenesis, organogenesis and organ development while genes whose products have signal transducer activity were enriched in florea and dorsata mushroom bodies compared to mellifera mushroom bodies. Probing these classes of genes in future studies might lead to a deeper understanding of the molecular basis of species differences in Apis.

In another promising result, genes that were upregulated in mellifera mushroom bodies compared to florea or dorsata were primarily involved in metabolism while genes enriched for catabolism were downregulated in florea and dorsata mushroom bodies compared to mellifera mushroom bodies. This result closely mirrors our earlier transcriptomic analysis of forager and one-day-old bees that also showed differences among these species in brain expression of metabolism genes [13]. Dyer [70] reported that mellifera colonies show higher rates of colony activity or "worker tempo" than florea or dorsata and have a higher colony metabolic rate. We speculate that to the extent that brain metabolism reflects whole organism metabolic activity our molecular results might in some way reflect these behavioral differences. Four genes involved in metabolism that showed species differences in both studies are alpha mannosidase (α-Man(II)b), Lethal (3) neo18, a serine-type carboxypeptidase (CG4678) and Ebony.

In addition to the four genes mentioned above, 34 other genes showed species differences in expression in our earlier study and species by CNS region differences in the present study. Some of the more obviously behaviorally related genes include orthologs of the Drosophila genes Doubletime (Dbt, also known as Discs overgrown), Synaptotagmin (Syt), Synaptotagmin IV (SytIV) and slowpoke. These genes are involved in circadian rhythms (Dbt, slowpoke) which figure prominently in dance behavior; [15] and synaptic activity and motor control (slowpoke and Synaptotagmins). They also provide good candidate genes to explore the molecular basis of dance language.

Acknowledgements

We would like to thank Raghavendra Gadagkar for extensive support during field collections; P. Kondaiah for essential infrastructural support; K.A. Ponnana for assistance in the field; Thomas Newman for laboratory support; Axel Brockmann for developing the RNALater ICE protocol for honey bee brains; Jenny Fell for help with array spot finding; Edwin Hadley for help with figures; and members of Robinson lab for reviewing this manuscript; and anonymous reviewers for comments that improved this manuscript.

Authors' Contributions

Conceived and designed the experiments: MSS GER. Performed the experiments: MSS. Analyzed the data: MSS SLRZ FH SZ. Contributed reagents/materials/analysis tools: GER. Wrote the paper: MSS GER.

References

1. Sunkin SM, Hohmann JG (2007) Insights from spatially mapped gene expression in the mouse brain. Hum Mol Genet 16(R2): R209–R219.

2. Hao Z, Ng F, Yixiao L, Hardin PE (2008) Spatial and circadian regulation of cry in Drosophila. J Biol Rhythms 23(4): 283–295.

3. Sillitoe RV, Joyner AL (2007) Morphology, molecular codes, and circuitry produce the three-dimensional complexity of the cerebellum. Annu Rev Cell Dev Biol 23: 549–577.

4. Hammock EA (2007) Gene regulation as a modulator of social preference in voles. Adv Genet 59: 107–127.

5. Mello CV, Velho TA, Pinaud R (2004) Song-induced gene expression: A window on song auditory processing and perception. Ann N Y Acad Sci 1016: 263–281.

6. Menzel R, Giurfa M (2001) Cognitive architecture of a mini-brain: The honeybee. Trends Cogn Sci 5(2): 62–71.

7. Giurfa M (2003) Cognitive neuroethology: Dissecting non-elemental learning in a honeybee brain. Curr Opin Neurobiol 13(6): 726–735.

8. Ismail N, Robinson GE, Fahrbach SE (2005) Stimulation of muscarinic receptors mimics experience-dependent plasticity in the honey bee brain. PNAS 103(1): 207–211.

9. Takeuchi H, Paul RK, Matsuzaka E, Kubo T (2007) EcR-A expression in the brain and ovary of the honeybee (Apis mellifera L.). Zool Sci 24(6): 596–603.

10. Paul RK, Takeuchi H, Kubo T (2006) Expression of two ecdysteroid-regulated genes, Broad-complex and E75, in the brain and ovary of the honeybee (Apis mellifera L.). Zool Sci 23(12): 1085–1092.

11. Mustard JA, Kurshan PT, Hamilton IS, Blenau W, Mercer AR (2005) Developmental expression of a tyramine receptor gene in the brain of the honey bee, Apis mellifera. J Comp Neurol 483(1): 66–75.

12. Whitfield CW, Ben-Shahar Y, Brillet C, Leoncini I, Crauser D, et al. (2006) Inaugural article: Genomic dissection of behavioral maturation in the honey bee. PNAS 103(44): 16068–16075.

13. Sen Sarma M, Whitfield CW, Robinson GE (2007) Species differences in brain gene expression profiles associated with adult behavioral maturation in honey bees. BMC Genomics 8: 202.

14. Grozinger CM, Fan Y, Hoover SE, Winston ML (2007) Genome-wide analysis reveals differences in brain gene expression patterns associated with caste and reproductive status in honey bees (Apis mellifera). Mol Ecol 16(22): 4837–4848.

15. Brockmann A, Robinson GE (2007) Central projections of sensory systems involved in honey bee dance language communication. Brain Behav Evol 70: 125–136.

16. Dyer FC (2002) The biology of the dance language. Annu Rev Entomol 47: 917–949.

17. Bausenwein B, Fischbach KF (1992) Activity labeling patterns in the medulla of Drosophila melanogaster caused by motion stimuli. Cell Tissue Res 270(1): 25–35.

18. Wiitanen W (1973) Some aspects of visual physiology of the honeybee. J Neurophysiol 36(6): 1080–1089.

19. Varela FG (1970) Fine structure of the visual system of the honey bee (Apis mellifera). II. the lamina. J Ultrastruct Res 31(1): 178–194.

20. Strausfeld NJ (2002) Organization of the honey bee mushroom body: Representation of the calyx within the vertical and gamma lobes. J Comp Neurol 450(1): 4–33.

21. Rybak J, Menzel R (1993) Anatomy of the mushroom bodies in the honey bee brain: The neuronal connections of the alpha-lobe. J Comp Neurol 334(3): 444–465.

22. Gronenberg W, Lopez-Riquelme GO (2004) Multisensory convergence in the mushroom bodies of ants and bees. Acta Biol Hung 55(1–4): 31–37.

23. Szyszka P, Galkin A, Menzel R (2008) Associative and non-associative plasticity in Kenyon cells of the honeybee mushroom body. Front Syst Neurosci 2(3): 10.3389/neuro.06.003.2008.

24. Okada R, Rybak J, Manz G, Menzel R (2007) Learning-related plasticity in PE1 and other mushroom body-extrinsic neurons in the honeybee brain. J Neurosci 27(43): 11736–11747.

25. Erber J (1978) Response characteristics and after effects of multimodal neurons in the mushroom body area of the honey bee. Physiol Entomol 3(2): 77–89.

26. Locatelli F, Bundrock G, Müller U (2005) Focal and temporal release of glutamate in the mushroom bodies improves olfactory memory in Apis mellifera. Journal of Neuroscience 25(50): 11614–11618.

27. Ehmer B, Gronenberg W (2002) Segregation of visual input to the mushroom bodies in the honeybee (Apis mellifera). J Comp Neurol 451(4): 362–373.

28. Lozano VC, Armengaud C, Gauthier M (2001) Memory impairment induced by cholinergic antagonists injected into the mushroom bodies of the honeybee. J Comp Physiol [A] 187(4): 249–254.

29. Komischke B, Sandoz JC, Malun D, Giurfa M (2005) Partial unilateral lesions of the mushroom bodies affect olfactory learning in honeybees Apis mellifera L. Eur J Neurosci 21(2): 477–485.

30. Homberg U (1985) Interneurones of the central complex in the bee brain (Apis mellifera, L.). J Insect Physiol 31(3): 251–261, 263–264.

31. Strauss R (2002) The central complex and the genetic dissection of locomotor behavior. Curr Opin Neurobiol 12(6): 633–638.

32. Wenzel B, Kunst M, Gunther C, Ganter GK, Lakes-Harlan R, et al. (2005) Nitric oxide/cyclic guanosine monophosphate signaling in the central complex of the grasshopper brain inhibits singing behavior. J Comp Neurol 488(2): 129–139.

33. Popov AV, Peresleni AI, Savvateeva-Popova EV, Wolf R, Heisenberg MER (2004) The role of the mushroom bodies and of the central complex of Drosophila melanogaster brain in the organization of courtship behavior and communicative sound production. J Evol Biochem Physiol 40(6): 641–652.

34. Homberg U (2008) Evolution of the central complex in the arthropod brain with respect to the visual system. Arth Struct & Dev 37(5): 347–362.

35. Sakura M, Lambrinos D, Labhart T (2008) Polarized skylight navigation in insects: Model and electrophysiology of e-vector coding by neurons in the central complex. J Neurophysiol 99(2): 667.

36. Snodgrass RE (1956) Anatomy of the honey bee. Ithaca, N.Y.: Comstock Pub Assoc.

37. Heinrich R, Wenzel B, Elsner N (2001) Pharmacological brain stimulation releases elaborate stridulatory behavior in gomphocerine grasshoppers–conclusions for the organization of the central nervous control. J Comp Physiol [A] 187(2): 155–169.

38. Heinrich R, Elsner N (1997) Central nervous control of hindleg coordination in stridulating grasshoppers. J Comp Physiol A 180(3): 257–269.

39. Anstey ML, Rogers SM, Ott SR, Burrows M, Simpson SJ (2009) Serotonin mediates behavioral gregarization underlying swarm formation in desert locusts. Science's STKE 323(5914): 627.

40. Honeybee Genome Sequencing Consortium (2006) Insights into social insects from the genome of the honeybee Apis mellifera. Nature 443(7114): 931–949.

41. Frisch Kv (1967) The dance language and orientation of bees. Cambridge, Massachusetts: Harvard University Press.

42. Kerr MK, Churchill GA (2001) Experimental design for gene expression microarrays. Biostatistics 2(2): 183–201.

43. ArrayExpress. [www.ebi.ac.uk_arrayexpress].

44. Whitfield CW, Cziko AM, Robinson GE (2003) Gene expression profiles in the brain predict behavior in individual honey bees. Science 302(5643): 296–299.

45. Cui X, Churchill GA (2003) Statistical tests for differential expression in cDNA microarray experiments. Genome Biol 4(4): 210.

46. Wu H, Kerr MK, Cui X, Churchill GA (2003) MAANOVA: A software package for the analysis of spotted cDNA microarray experiments. In: Parmigiani GG, editor. The Analysis of Gene Expression Data: Methods and Software. Springer Verlag.

47. Benjamini Y, Hochberg Y (1995) Controlling the false discovery rate: A practical and powerful approach to multiple testing. Journal -Royal Statistical Society Series B 57: 289–289.

48. Kamikouchi A, Takeuchi H, Sawata M, Ohashi K, Natori S, et al. (1998) Preferential expression of the gene for a putative inositol 1, 4, 5-trisphosphate receptor homologue in the mushroom bodies of the brain of the worker honeybee Apis mellifera L. Biochem Biophys Res Commun 242(1): 181–186.

49. Smith LA, Peixoto AA, Kramer EM, Villella A, Hall JC (1998) Courtship and visual defects of cacophony mutants reveal functional complexity of a calcium-channel α1 subunit in Drosophila. Genetics 149(3): 1407–1426.

50. Kamikouchi A, Takeuchi H, Sawata M, Natori S, Kubo T (2000) Concentrated expression of Ca+2/Calmodulin-dependent protein kinase II and protein kinase C in the mushroom bodies of the brain of the honeybee Apis mellifera L. J Comp Neurol 417: 501–510.

51. Martinek S, Inonog S, Manoukian AS, Young MW (2001) A role for the segment polarity gene shaggy/GSK-3 in the Drosophila circadian clock. Cell 105(6): 769–779.

52. Michel M, Kemenes I, Muller U, Kemenes G (2008) Different phases of long-term memory require distinct temporal patterns of PKA activity after single-trial classical conditioning. Learn Mem 15(9): 694–702.

53. Müller U (2000) Prolonged activation of cAMP-dependent protein kinase during conditioning induces long-term memory in honeybees. Neuron 27(1): 159–168.

54. Muller U (1999) Second messenger pathways in the honeybee brain: Immunohistochemistry of protein kinase A and protein kinase C. Microsc Res Tech 45(3): 165–173.

55. Joiner MA, Griffith LC (1997) CaM kinase II and visual input modulate memory formation in the neuronal circuit controlling courtship conditioning. J Neurosc 17(23): 9384–9391.

56. Pascual A, Preat T (2001) Localization of long-term memory within the Drosophila mushroom body. Science 294(5544): 1115–1117.

57. Kiya T, Kunieda T, Kubo T (2007) Increased neural activity of a mushroom body neuron subtype in the brains of forager honeybees. PLoS ONE 2(4): e371.

58. Okada R, Sakura M, Mizunami M (2003) Distribution of dendrites of descending neurons and its implications for the basic organization of the cockroach brain. J Comp Neurol 458(2): 158–174.

59. Farris SM (2008) Structural, functional and developmental convergence of the insect mushroom bodies with higher brain centers of vertebrates. Brain Behav Evol 72(1): 1–15.

60. Han PL, Meller V, Davis RL (1996) The Drosophila brain revisited by enhancer detection. J Neurobiol 31(1): 88–102.

61. Roeder T, Schramm G, Marquardt H, Bussmeyer I, Franz O (2004) Differential transcription in defined parts of the insect brain: Comparative study utilizing Drosophila melanogaster and Schistocerca gregaria. Invertebr Neurosci 5(2): 77–83.

62. Spaethe J, Chittka L (2003) Interindividual variation of eye optics and single object resolution in bumblebees. J Exp Biol 206(Pt 19): 3447–3453.

63. Jander U, Jander R (2002) Allometry and resolution of bee eyes (Apoidea). Arth Struct & Dev 30(3): 179–193.

64. Oldroyd BP, Wongsiri S (2006) Asian honey bees : Biology, conservation, and human interactions. Cambridge, Massachusetts: Harvard University Press.

65. Cash AC, Whitfield CW, Ismail N, Robinson GE (2005) Behavior and the limits of genomic plasticity: Power and replicability in microarray analysis of honeybee brains. Genes, Brain and Behavior 4(4): 267–271.

66. Brockmann A, Annangudi SP, Richmond TA, Ament SA, Xie F, et al. (2009) Quantitative peptidomics reveal brain peptide signatures of behavior. Proceedings of the National Academy of Sciences 106(7): 2383–2388.

67. Wessnitzer J, Webb B (2006) Multimodal sensory integration in insects—towards insect brain control architectures. Bioinspir Biomim 1: 63–75.

68. Arias MC, Sheppard WS (2005) Phylogenetic relationships of honey bees (Hymenoptera:Apinae:Apini) inferred from nuclear and mitochondrial DNA sequence data. Mol Phylogenet Evol 37(1): 25–35.

69. Raffiudin R, Crozier RH (2007) Phylogenetic analysis of honey bee behavioral evolution. Mol Phylogenet Evol 43(2): 543–552.

70. Dyer FC, Seeley TD (1991) Nesting behavior and the evolution of worker tempo in four honey bee species. Ecology 78: 156–170.

71. Brandt R, Rohlfing T, Rybak J, Krofczik S, Maye A, et al. (2005) Three-dimensional average-shape atlas of the honeybee brain and its applications. J Comp Neurol 492(1): 1–19.

CITATION

Plant Volatiles, Rather than Light, Determine the Nocturnal Behavior of a Caterpillar

Kaori Shiojiri, Rika Ozawa and Junji Takabayashi

ABSTRACT

Although many organisms show daily rhythms in their activity patterns, the mechanistic causes of these patterns are poorly understood. Here we show that host plant volatiles affect the nocturnal behavior of the caterpillar Mythimna separata. Irrespective of light status, the caterpillars behaved as if they were in the dark when exposed to volatiles emitted from host plants (either uninfested or infested by conspecific larvae) in the dark. Likewise, irrespective of light status, the caterpillars behaved as if they were in the light when exposed to volatiles emitted from plants in the light. Caterpillars apparently utilize plant volatile information to sense their environment and modulate their daily activity patterns, thereby potentially avoiding the threat of parasitism.

Introduction

Photoperiod and the temperature are the most stable abiotic rhythms on the earth, and they are considered to be the principal exogenous factors that affect daily periodicity of behaviors of all organisms, particularly nocturnal feeders [1]. Nonetheless, many cues, including biotic factors, are available to foraging organisms, and each cue may have different potentials in determining daily activity patterns. For example, biotic factors such as predation or foraging ability based on sensory modalities would influence diel activity patterns. What is not well understood is how abiotic and biotic factors act in concert to regulate the daily activity patterns of foraging organisms. For example, studies of the effects of photoperiod on herbivorous insects typically place the insects on host plants, and behaviors are observed during different photoperiods. However, in addition to the light–dark cycle, the host plant is also potentially influencing the behavior of the insect because host plants are often not only food sources for herbivorous arthropods, but also their microhabitats. Host plants release odors, or volatiles, that have been shown to be influenced by photoperiod [2]. Thus, herbivorous arthropods are also confronted with a number of plant factors that potentially vary according to photoperiod and temperature, and it is critical to understand the role of each of these factors when determining the mechanisms regulating diel activity patterns.

Host plants are known to emit specific blends of volatiles in response to herbivory, and such volatiles are called herbivore-induced plant volatiles (HIPV) [3, 4]. Interestingly, HIPV show diurnal patterns [2, 5, 6]. For example, corn plants infested by larvae of the noctuid Spodoptera exigua emit S. exigua–induced plant volatiles (S. exigua–IPV) that attract parasitic wasps [7]. S. exigua–IPV are composed of several monoterpenoids, sesquiterpenoids, green leaf volatiles, and the compound indole [7]. The production of S. exigua–IPV shows daily periodicity; emission increases in the daytime and decreases in the nighttime [2, 5]. Cotesia marginiventris, parasitic wasps of S. exigua larvae, are attracted to S. exigua–IPV in wind tunnel experiments [8], suggesting that the wasps actively search for S. exigua larvae during the day using S. exigua–IPV in the field. Maeda et al. (2000) also reported similar patterns in a tritrophic system consisting of kidney bean plants, the herbivorous mite Tetranychus urticae, and predatory mites Phytoseiulus persimilis: the production of T. urticae–IPV that attract the predatory mites increases during the day and decreases in the night [6]. These data suggest that host plants during the nighttime are an enemy-free space due to the lack of HIPV production, and such diurnal changes in HIPV production may play an important role in determining the day–night patterns of herbivorous insects.

Mythimna separata (Lepidoptera: Noctuidae) is a caterpillar that feeds on many graminaceous plant species. Corn plants infested by M. separata larvae emit

M. separata–IPV that attract the parasitic wasp Cotesia kariyai (Hymenoptera: Braconidae) [9]. Sato et al. (1983) reported that C. kariyai was diurnal whereas M. separata larvae fed and were active primarily at night and are thus nocturnal [10]. When studying the day–night patterns of M. separata larvae, they used potted corn plants as food and observed the feeding and hiding behavior of the larvae on these plants [10]. However, the behavior of the larvae might have been affected by the day–night patterns in the production of infested corn plant volatiles, and they did not investigate this potentially confounding factor. Here we show for the first time that differences in volatiles from corn plants (either uninfested or infested by M. separata larvae) under light and dark conditions are critical factors affecting the daily periodicity of this herbivorous insect. In fact, we show that volatile cues were more important than light cues in modulating the hiding behavior of M. separata. We discuss possible explanations for the evolutionary and ecological significance of host plant volatiles as diel cues in a tritrophic context.

Results

Effects of Light on the Hiding Behavior of the Larvae

When offered only artificial diet, the numbers of larvae exhibiting hiding behavior were not different under the two light conditions (Figure 1, black lines: generalized linear model, $p = 0.754$). This suggests that light alone was not sufficient to affect the hiding behavior of the larvae.

Effects of the Presence of Plants on the Hiding Behavior of the Larvae

The presence of plants affected caterpillar hiding behavior either positively or negatively relative to caterpillars without plants (generalized linear model, $p < 0.0001$). The interaction between light conditions and plant factors was also significant (generalized linear model, $p < 0.0001$).

Under the daytime light condition, the number of hiding larvae in the experiment with corn plants nearby was 12% (at 2 h), 21% (4 h), 25% (6 h), and 19% (8 h) higher than without corn plants (Fisher's exact probability test, at 2 h: $p = 0.15$, 4 h: $p = 0.01$, 6 h: $p = 0.003$, and 8 h: $p = 0.03$; Figure 1, black dashed line and green dashed line). Under dark conditions, on the other hand, the number of hiding larvae in the experiment with corn plants nearby was 14% (at 2 h), 35% (4 h), 30% (6 h), and 32% (8 h) lower than without corn plants (Fisher's exact probability test, at 2 h: $p = 0.126$, 4 h: $p = 0.0002$, 6h: $p = 0.003$, and 8 h: $p = 0.0012$; Figure 1, black line and green line).

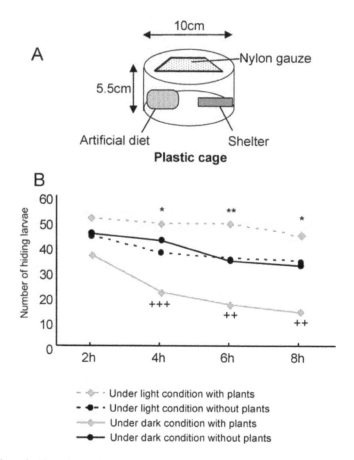

Figure 1. Effects of Light and Uninfested Corn Plants on the Hiding Behavior of M. separata

(A) An illustration of the plastic cup in which the individual larvae are placed during the experiment.

(B) The graph shows the number of larvae hiding (y-axis) in 8 h (x-axis).

Asterisks (*) indicate the comparison between dashed lines (i.e., larvae under light with or without plants): A single asterisk (*) indicates 0.01 < p < 0.05, and double asterisks (**) indicate 0.001 < p < 0.01, by Fisher's exact probability test.

Plus signs (+) indicate the comparison between solid lines (i.e., larvae under dark with or without plants): double plus signs (++) indicate 0.001 < p <0.01, and triple plus signs (+++) indicate p < 0.001, by Fisher's exact probability test.

Effects of the Plant Volatiles on the Hiding Behavior of the Larvae

The experimental setups are illustrated in Figure 2A. The light conditions of the plants affected hiding behavior of the larvae (generalized linear model, Figure 2B: p < 0.0001 and Figure 2C: p < 0.0001). The light conditions of the larvae (generalized linear model, Figure 2B: p = 0.809 and Figure 2C: p = 0.416) and the interactions between the light conditions of the plants and larvae (generalized linear model, Figure 2B: p = 0.9391 and Figure 2C: p = 0.4722) were not significant.

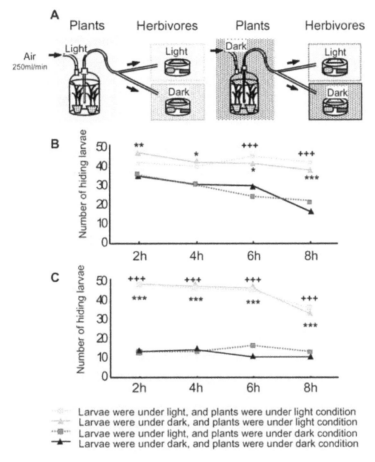

Figure 2. Effects of Light and the Volatiles from Plants under Either Light or Dark Conditions on the Hiding Behavior of M. separata

(A) Experimental setups. Volatiles from corn plants were split into two airstreams and sent to incubators in which the larvae were kept under either dark or light conditions.

(B) Number of larvae hiding when exposed to uninfested corn plant volatiles.

(C) Number of larvae hiding when exposed to infested corn plant volatiles.

For (B) and (C), asterisks (*) indicate the comparison between the blue line and the black line (i.e., the comparison of the effects of volatiles from uninfested/infested plants under light with those under dark on the hiding behavior of the larvae under dark): A single asterisk (*) indicates 0.01< p <0.05, double asterisks (**) indicate 0.001 < p < 0.01, and triple asterisks (***) indicate p < 0.001, by Fisher's exact probability test. Plus signs (+) indicate the comparison between the green line and the red line (i.e., the comparison of the effects of volatiles from uninfested/infested plants under light with those under dark on the hiding behavior of the larvae under light): Triple plus signs (+++) indicate p < 0.001, by Fisher's exact probability test.

When larvae were in the dark, the number of larvae hiding when experiencing volatiles from uninfested corn plants in light was 24% (at 2 h), 22% (4 h), 22% (6 h), and 42% (8 h) higher than those experiencing volatiles from uninfested corn plants in the dark (Fisher's exact probability test, at 2 h: p= 0.005, 4 h: p = 0.027, 6 h: p = 0.030, and 8 h: p < 0.0001; Figure 2B). Likewise, when larvae were in the light, the number of larvae hiding when experiencing volatiles from uninfested corn plants in light was 12% (at 2h), 18% (4h), 40% (6h), and 38% (8 h) higher than those experiencing volatiles from uninfested corn plants in the dark (Fisher's exact probability test, at 2 h: p = 0.24, 4 h: p = 0.083, 6 h: p < 0.0001, and 8 h: p < 0.0001; Figure 2B). These data indicate that it is not the light condition experienced by the larvae, but the volatiles from the corn plants (either under light or dark conditions) that affect the hiding behavior of the caterpillars.

The effects of volatiles on the larvae were similar on infested and uninfested plants. Irrespective of the light condition of the larvae, the number of larvae that hid when experiencing volatiles from infested corn plants in light was higher than larvae experiencing volatiles from infested corn plants in the dark (larvae under dark conditions: at 2 h (68%), 4 h (64%), 6 h (70%), and 8 h (44%); p < 0.0001 for each observation time; larvae under light conditions: at 2 h (68%), 4 h (64%), 6 h (56%), and 8 h (44%); p < 0.0001 for each observation time; Figure 2C).

Discussion

Studies of day–night (or light–dark) patterns of herbivorous insects typically place the insects on host plants, and their behaviors are observed during different photoperiods. However, in this commonly used experimental design, light condition is confounded with other plant factors such as the emission of host plant volatiles. Sato et al. (1983) observed day–night behavioral patterns of M. separata larva on potted corn plants: The larva hid in a sheath of the host plant or underground during the day, and fed on the plants during the night [10]. Based on these data, they concluded that M. separata was nocturnal. However, while we were rearing M. separata larvae on an artificial diet, we found that they did not exhibit this day–night behavioral pattern (Figure 1), suggesting that factors other than light could be important in determining the nocturnal behaviors that Sato et al. (1983) observed [10]. One important difference was that we used an artificial diet whereas Sato et al. (1983) used corn plants as food [10]. As clearly seen in Figure 1, factors from uninfested corn plants in the light enhance the daytime behavior of hiding in a shelter for M. separata larvae, whereas factors from plants in the dark enhance the nighttime behavior of leaving the shelter and feeding.

As corn plants show diurnal variation in the production of volatiles [5], we hypothesized that a difference in the blend of volatiles from corn plants that depends on light conditions affects the differences between daytime and nighttime behaviors of M. separata larvae. To test this hypothesis, we first provided volatiles from uninfested corn plants that were under either light or dark conditions, to the group of larvae in cups that were under either light or dark conditions (Figure 2B). The data indicate that it is not the light condition of the larvae, but rather the volatiles from the uninfested corn plants that affect hiding behavior. We then repeated the experiments using corn plants infested by M. separata larvae as a source of volatiles. The larvae showed the same behavior as seen in the experiments using uninfested plants as an odor source: Irrespective of light conditions of the larvae, significantly more larvae hid when volatiles from infested plants under light conditions were offered than when volatiles under dark conditions were offered (Figure 2C). It is important to note that the difference was more pronounced with volatiles from the infested plants (Figure 2C). Plant volatiles are known to have numerous functions, such as attracting herbivores [11, 12] and/or carnivorous natural enemies of herbivores [3, 4, 7], repelling herbivores [13, 14], and attracting pollinators [15, 16], to name a few. Here, we suggest an additional function: Plant volatiles may affect the diurnal and nocturnal behavior of herbivores.

The circadian rhythms of insects have long been studied, and there are several potential causes for circadian rhythms [1]. We believe that the use of changes in host plant volatiles to alter diel patterns of feeding fits within a tritrophic framework. Insect parasitoids typically forage during the day [10, 17] and use diurnal plant volatile emissions as foraging cues [2, 5]. Consequently, caterpillars using day–night changes in the profiles of host plant volatiles to regulate their own activity rhythms could reduce the probability of parasitism by exploiting temporally available enemy-free space. Additional experiments are planned to determine (1) the compounds of plant volatiles that significantly affect the nocturnal behaviors of M. separata larvae either positively or negatively, and (2) if the modification of diel activity cycles by host plant volatiles is a widespread phenomenon in herbivorous insects, especially in those species that leave their host plants when not feeding.

Materials and Methods

Insects and Plants

M. separata was obtained from a culture reared at Tsukuba University in Tsukuba, Ibaraki, Japan. The insects were reared in our laboratory on artificial diet (Insecta LF, Nihon Nousan Kogyo, Yokohama, Japan) under conditions of 25 ± 2 °C, 24-h dark, 50%–70% relative humidity. This photoperiodic regime was selected to ensure that there would be no photoperiodically entrained diel periodicity at the time the different assays were carried out.

Potted corn plants (Zea mays L. cv. Royal Dent) (three plants per pot) were grown in a growth chamber (25 ± 2 °C, 16:8-h light:dark), and 10-d-old plants were used in all experiments.

Effects of Light on the Hiding Behavior of the Larvae

Based on the data by Sato et al. (1983), M. separata larvae feed and are active primarily at night [10]. They hide in the sheath of a host plant or under the ground during the day. Here, we focused on hiding as the criterion for diurnal behavior. We placed 57 third-stadium larvae in individual plastic cups (20-cm diameter and 6-cm height) with filter paper shelters (folded in accordion manner: 4 × 4 cm), each with a piece of artificial diet (ca. 7 g). The top of each cup had a 5 × 5 cm nylon gauze window (Figure 1A). The cups were kept in an incubator under either 6500 lux light conditions or dark conditions for 8 h at 25 ± 2 °C with 50%–70% relative humidity. We observed larval behavior every 2 h, to determine whether or not larvae were hiding in the shelters.

Effects of the Presence of Plants on the Hiding Behavior of the Larvae

To study the effects of the host plants on the feeding behavior of the larvae, we conducted similar experiments as above, only with uninfested corn plants. Uninfested corn plants are those that have never experienced herbivory. We placed six pots of uninfested corn plants, each containing three plants of circa 30 cm height per pot, around the 57 cups containing third-stadium larvae in an incubator of the same climate conditions as above.

Effects of the Plant Volatiles on the Hiding Behavior of the Larvae

To test the hypothesis that host plant volatiles affect the diurnal feeding behavior of the caterpillars, we conducted similar experiments as above, with the addition of plant volatiles. Four pots of three plants each were placed into a 7,200-ml plastic separable flask. Volatile flow emitted from the plants (under either light [6,500 lux] or dark conditions) were collected in a 250-ml/min flow stream that was split into two, and each flow was directed to an incubator containing 50 third-stadium larvae. These larvae were housed individually in cups for 8 h under either light

(6500 lux) or dark conditions (Figure 2; 25 ± 2 °C, 50%–70% relative humidity). We checked whether the larvae were hiding in the shelters every 2 h. We used both intact and infested corn plants as the odor source. To obtain infested plants, we placed ten third-stadium M. separata larvae in each pot on the corn plants. After 18 h, the larvae and their feces were removed, leaving behind a damaged area of circa 10% of the total leaf surface. In these experiments, all caterpillars were subjected to the same airflows; however, we independently tested day and night volatiles from infested and uninfested plants, and obtained similar results.

Statistics

We analyzed the data in Figure 1 using generalized linear models and binomial errors with the software program R 1.7.0 for Windows [18] in order to test the hypothesis that it was not the light condition experienced by the larvae, but factors from intact corn plants that affected the hiding behavior of the caterpillars. Differences in hiding behaviors between different conditions of larvae (e.g., light conditions of larvae, and presence or absence of intact plants nearby) were analyzed. We also analyzed the data in Figure 2A and 2B with R to test the hypothesis that it was the volatiles from corn plants that affected the hiding behavior of the larvae and not the light condition experienced by the larvae. Differences in hiding behavior between different conditions of larvae (e.g., light conditions of larvae and the light condition of the plants) were analyzed. Then, the numbers of larvae hiding in different treatments were compared with Fisher's exact probability test.

Acknowledgements

We thank R. Karban for comments on the manuscript, and G. Takimoto and M. Uefune for comments on statistical analysis.

Authors' Contributions

JT conceived and designed the experiments. KS performed the experiments. RO analyzed the data.

References

1. Saunders DS (2002) Insect Clocks. 3rd edition. Amsterdam: Elsevier Science. 576 p.

2. Loughrin JH, Manukian A, Heath RR, Turlings TCJ, Tumlinson JH (1994) Diurnal cycle of emission of induced volatile terpenoids by herbivore-injured cotton plants. Proc Natl Acad Sci USA 91: 11836–11840.

3. Takabayashi J, Dicke M (1996) Plant-carnivore mutualism through herbivore-induced carnivore attractants. Trends Plant Sci 1: 109–113.

4. Dicke M, Vet LEM (1999) Plant-carnivore interactions: Evolutionary and ecological consequences for plant, herbivore and carnivore. In: Olff H, Brown VK, Drent RH, editors. Herbivores: Between plants and predators. Oxford (United Kingdom): Blackwell Science. pp. 483–520.

5. Turlings TCJ, Loughrin LJ, McCall PJ, Rose US, Lewis WJ, et al. (1995) How caterpillar-damaged plants protect themselves by attracting parasitic wasps. Proc Natl Acad Sci USA 92: 4169–4174.

6. Maeda T, Takabayashi J, Yano S, Takafuji A (2000) Effects of light on the tritrophic interaction between kidney bean plants, two-spotted spider mites and predatory mites, Amblysieus womersleyi (Acari: phytoseiidae). Exp Appl Acarol 24: 415–425.

7. Turlings TCJ, Tumlinson JH, Lewis WJ (1990) Exploitation of herbivore-induced plant odors by host-seeking parasitic wasps. Science 250: 1251–1253.

8. Turlings TCJ, Fritzsche ME (1999) Attraction of parasitic wasps by caterpillar-damaged plants. Novartis Foundation. Insect-plant interactions and induced plant defense, No. 223. John Wiley & sons. pp. 21–38. Available at: http://www.wiley.com/WileyCDA/WileyTitle/productCd-0471988154.html. Accessed 4 April 2006.

9. Takabayashi J, Takahashi S, Dicke M, Posthumus MA (1995) Developmental stage of the herbivore Pseudaletia separata affects production of herbivore-induced synomone by corn plants. J Chem Ecol 21: 273–287.

10. Sato Y, Tanaka T, Imafuku M, Hidaka T (1983) How does diurnal Apanteles kariyai parasitize and edress from nocturnal host larva? Kontyu 51: 128–139.

11. Reddy GVP, Guerrero A (2004) Interactions of insect pheromones and plant semiochemicals. Trend Plant Sci 9: 253–261.

12. Landolt PJ, Tumlinson JH, Alborn DH (1999) Attraction of Colorado potato beetle (Coleoptera: Chrysomelidae) to damaged and chemically induced potato plants. Environ Entomol 28: 973–978.

13. Gibson RW, Pickett JA (1983) Wild potato repels aphids by release of aphid alarm pheromone. Nature 302: 608–609.

14. De Moraes CM, Mescher MC, Tumlinson JH (2001) Caterpillar-induced nocturnal plant volatiles repel conspecific females. Nature 410: 577–580.

15. Andersson S, Dobson HEM (2003) Antennal responses to floral scents in the butterfly Heliconius melpomene. J Chem Ecol 29: 2319–2330.

16. Terry I, Moore CJ, Walter GH, Forster PI, Roemer RB, et al. (2004) Association of cone thermogenesis and volatiles with pollinator specificity in Macrozamia cycads. Plant Syst Evol 243: 233–247.

17. Quicke DLJ (1997) Parasitic wasps. 1st edition. London: Chapman & Hall. 470 p.

18. Ihaka R, Gentleman R (1996) R: A language for data analysis and graphics. J Comput Graph Analysis 5: 299–314.

CITATION

Risk and Ethical Concerns of Hunting Male Elephant: Behavioral and Physiological Assays of the Remaining Elephants

Tarryne Burke, Bruce Page, Gus Van Dyk, Josh Millspaugh
and Rob Slotow

ABSTRACT

Background

Hunting of male African elephants may pose ethical and risk concerns, particularly given their status as a charismatic species of high touristic value, yet which are capable of both killing people and damaging infrastructure.

Methodology/Principal Findings

We quantified the effect of hunts of male elephants on (1) risk of attack or damage (11 hunts), and (2) behavioral (movement dynamics) and physiological (stress hormone metabolite concentrations) responses (4 hunts) in Pilanesberg National Park. For eleven hunts, there were no subsequent attacks on people or infrastructure, and elephants did not break out of the fenced reserve. For three focal hunts, there was an initial flight response by bulls present at the hunting site, but their movements stabilised the day after the hunt event. Animals not present at the hunt (both bulls and herds) did not show movement responses. Physiologically, hunting elephant bulls increased faecal stress hormone levels (corticosterone metabolites) in both those bulls that were present at the hunts (for up to four days post-hunt) and in the broader bull and breeding herd population (for up to one month post-hunt).

Conclusions/Significance

As all responses were relatively minor, hunting male elephants is ethically acceptable when considering effects on the remaining elephant population; however bulls should be hunted when alone. Hunting is feasible in relatively small enclosed reserves without major risk of attack, damage, or breakout. Physiological stress assays were more effective than behavioral responses in detecting effects of human intervention. Similar studies should evaluate intervention consequences, inform and improve best practice, and should be widely applied by management agencies.

Introduction

Successful lobbying against hunting practices by animal-welfare and animal-rights groups [1] as well as limited data regarding the potentially negative long-term effects of direct intervention activities [2] has generated public concern around effects of management intervention on animal species. This is especially true for species that hold special appeal to humans in terms of their charisma, size, danger and drama associated with them [3], [4], and even more so if these same species appear on rare and/or endangered lists [3]. Management interventions that are perceived negatively also have the potential to reduce public appeal, and hence tourism, not only to specific reserves but to protected areas in general. In Africa in particular, many protected areas are dependent on the revenue generated by non-consumptive tourism [5], which therefore has implications for the continuation and very survival of its protected areas.

African elephants (Loxodonta africana) hold special appeal to humans not only due to their high tourist value as one of the 'Big 5' species, but also because they are social animals that form strong and long-lasting bonds between individuals [6], resulting in humans developing a strong sense of empathy for them. Further, large animals such as elephants pose a potential danger to human life, to infrastructure, and can break out of the fenced areas in which they occur. Major management intervention such as hunting may elicit unpredictable, dangerous responses, and management agencies have to minimise such risks. From both ethical and conservation perspectives, it is therefore essential to quantify the effects of direct human intervention on animal populations.

Animal physiological stress can be measured non-invasively through the measurement of glucocorticoid metabolites (i.e., cortisol or corticosterone) in faeces across a variety of taxa (e.g. 7. 8. [9]). This allows for an accurate assessment of stress without the bias of capture- or disturbance-induced increases in glucocorticoid levels (e.g. [8], [10], [11]). Stress assessment in wildlife serves as a forewarning of possible deleterious impacts from human activities [12].

We therefore aimed to determine the effect of direct human intervention on elephants in a small reserve through behavioral observations and the quantification of the physiological stress responses using faecal hormone metabolite concentrations. Specifically, we determined whether, in response to hunting, elephant showed changes in their (1) movement patterns, (2) grouping patterns (among the breeding herds), and (3) physiological stress levels. Further, we assessed whether physiological stress assays and behavioral observations corresponded. Most importantly, we assessed for extreme responses on the part of the elephants such as attacks on people, infrastructure, or breaking out from the reserve.

Results

'Major' Events

BE01 (BE = Bull elephant) was hunted in June 1996, was wounded by the client, and followed-up by the support-team. He charged, and was shot by the Professional Hunter, but managed to kill the Professional Hunter before collapsing. Apart from this, no other major incidents occurred as a result of any hunts, either among the remaining bulls or the breeding herds. Elephants did not break out of the Park or cause damage to infrastructure, and were not responsible for any tourist-related incidents relative to any of the eleven bull hunt events that occurred in the Park.

Responses of Individuals Present at the Hunt of the Targeted Bull

For the three hunt events where bulls were present, in all cases the bulls rapidly moved away from the hunt site, and displaced a relatively large distance. In general, the bulls that were present increased their distance from the hunt site in the ten days following the hunt relative to the ten day prior to the hunt. There was no pattern for a change in the direction of movement of bulls associated with the hunted animal in the ten days before relative to after the hunts. For one hunt, the two bulls present at the hunt increased their daily displacement rate after the hunt, but this did not change for bulls present at the two other hunts. There was no clear pattern of shifting home range for those individuals present at the hunt event.

Physiologically, the maximum time taken for those bulls that were in the presence of the hunted bull to return to levels of faecal stress hormone metabolite concentrations similar to their baseline levels in the one-month period prior to the hunt event was four days, i.e. there was a clear, relatively short, physiological response (Figure 1). To assess a slightly longer physiological response, we used a before-after control design. However, we factored out the four days after the hunt event to remove the extreme response indicated above. We thus compared the average stress levels over a one-month period, and compared the one month before the hunt to the one month starting five days after the hunt. There was no significant difference between the 'one-month before hunt' average faecal stress hormone metabolite concentration and the 'one-month after hunt' average faecal stress hormone metabolite concentration (Wilcoxon signed-ranks test: $T = -0.11$, $N = 6$, $P = 0.92$). There was a significant increase in the six individuals' (that were present at the hunt events) average maximum (of the three maximum values) faecal stress hormone metabolite concentration in the four-day period following the hunt relative to their 'one-month before hunt' baseline metabolite concentration values (Wilcoxon signed-ranks test: $T = -2.20$, $N = 6$, $P = 0.028$) (Figure 1). The hunt events therefore induced significant physiological stress responses in those individuals that were present at the actual hunting of the targeted individual. The behavioral observations of these same bulls showed that they exhibited a 'flight' response to the hunt events, but that their movements stabilised after one day following the hunt. Thus the physiological stress response was more long-lived than the behavioral 'flight' response.

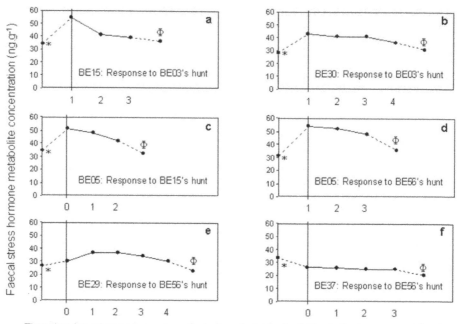

Figure 1. The effect of bull hunts on other bulls present at the hunt. We present the faecal stress hormone metabolite concentrations of the six individual bulls that were in association with the respective targeted bulls at the time of their hunts. '*' represents the individual's baseline stress hormone concentrations in the one-month period before the respective hunt event; 'Φ' represents the individual's baseline stress hormone concentrations in the one-month period after the respective hunt event. Vertical lines represent the first day relative to the hunt events when faecal samples from individuals were collected (0 = day of hunt, 1 = one day after hunt etc.). Up to four days passed after the hunt events before the individuals' faecal stress hormone metabolite concentrations were comparable with their initial (one month before hunt) baseline values. There was a significant increase between the maximum faecal stress hormone metabolite concentration and the respective 'before hunt' baseline faecal stress hormone metabolite concentrations (Wilcoxon signed-ranks test: T = –2.20, N = 6, P = 0.028).

Response of Individuals not in Close Proximity to the Hunt of the Targeted Bull

None of the four hunt events induced significant changes in the direction of movement relative to the respective hunt sites for any of the bulls or breeding herds not in close proximity to the hunt. (Table 1). Of the total number of bulls analysed, only 11% were found to move significantly further and 7% to move significantly closer to the respective hunt sites in the ten-day period following the hunt events (Table 1). For the breeding herds, 45% moved significantly further and 7% significantly closer to the respective hunt sites in the ten-day period following the hunt. (Table 1). Forty three percent of all the bulls analysed relative to the four hunt events showed an increase in their displacement rates following the hunts (Table 1); thus approximately half of the bulls increased and half decreased their displacement rates following bull hunt events. Similarly, 45% of all of the breeding herds analysed showed an increase in their displacement rates following the hunt events (Table 1); thus approximately half of the herds increased and half decreased their displacement rates in response to bull hunt events.

Table 1. Effects of four hunts on the movement dynamics of bulls and breeding herds not present at the hunts of the targeted bulls.

Sex	Hunt	Number assessed	Move further	Move closer	Faster rate	Slower rate	Increase Range	Decrease Range
Bulls	BE03	4	0	2	1	3	2	2
	BE15	7	1	0	1	6	4	3
	BE56	7	1	0	3	4	1	6
	BE28	10	1	0	6	4	5	5
Breeding Herds	BE03	4	3	0	3	1	1	3
	BE15	4	2	0	2	2	1	3
	BE56	10	8	0	5	5	7	3
	BE28	11	0	2	7	4	4	7

Data are the number of bulls or breeding herds that responded out of those for which we had data for that particular hunt (number assessed).
Only those bulls and breeding herds that showed significant changes in response variables (Mann-Whitney: p<0.05) are presented (i.e., numbers in some columns do not add up to number assessed). No animals showed a directional shift towards or away from the hunts, and those response variables are not included in the table.

Table 2. The effects of the four hunts on bulls' and breeding herds' ranging from one month before to one month after the respective hunt events

Sex	Hunt	N	Average (%) Overlap in Total Range after vs before [a]	Core Range Average Increase Factor	Individuals with increase in core range size (%) [b]
Bulls	BE03	6	83.5	1.7	67
Bulls	BE56	10	70.2	7.3	70
Bulls	BE15	9	66.4	2.3	56
Bulls	BE28	10	61.8	3.1	50
Breeding herds	BE03	4	81.8	2.6	75
Breeding herds	BE56	6	79.7	5.6	67
Breeding herds	BE15	5	72.6	2.5	20
Breeding herds	BE28	7	42.3	2.8	14

[a]Percent overlap of the total range (i.e. area enclosed by 95% Kernel) after hunts to the range before the hunts.
[b]Percent of individuals whose core home ranges (i.e. areas enclosed by the 50% Kernel) increased from 'before' to 'after' the hunt events (N is given in third column).

There was no significant change in the bulls' core home range sizes in response to the hunt in three cases (Wilcoxon signed-ranks test: Hunt of BE03: $T = -0.31$, $N = 6$, $P = 0.75$; Hunt of BE15: $T = -0.56$, $N = 9$, $P = 0.58$; BE28: $T = -0.36$, $N = 10$, $P = 0.72$), while there was a significant increase in the bulls' core home range sizes in the month after the hunt of BE56 occurred ($T = -2.09$, $N = 10$, $P = 0.037$) (Table 2).

There was no significant change in the respective breeding herds' core home range sizes in response to the hunt in three cases (Wilcoxon signed-ranks test: BE03: $T = -1.46$, $N = 4$, $P = 0.14$; BE15: $T = -1.75$, $N = 5$, $P = 0.08$; BE28: $T = -1.42$, $N = 11$, $P = 0.16$) (Table 2). As in the case of the bulls, there was a significant increase in the breeding herds' core home range sizes in response to the hunt of BE56 ($T = -2.29$, $N = 10$, $P = 0.022$) (Table 2).

For the animals not present at the hunt, we integrated the analysis of behavioral responses using a combined probabilities test. Bull hunt events did not have substantial impacts on the remaining elephants' movement dynamics, bulls' and breeding herds' distances from the hunt sites, direction of movements relative to the hunt sites, displacement rates, core home range sizes, and, for the breeding herds, the fission and fusion dynamics (P>0.05 for all tests) (Table 3).

Table 3. Overall effects of the four bull hunts on all the adult bulls and independent breeding herds analysed using combined probabilities tests (Sokal and Rohlf, 1981).

Sex	Response	Statistical Test	$-2\sum$ln P	d.f.	P
Bulls	Distance from hunt site	Mann-Whitney	40.55	70 [a]	>0.995
	Direction of movement	Sign	31.40	70 [a]	>0.999
	Rate of movement	Wilcoxon	3.72	8 [b]	>0.75
	Core home range size	Wilcoxon	3.03	8 [b]	>0.90
Cows	Distance from hunt site	Mann-Whitney	63.16	60 [c]	>0.25
	Direction of movement	Sign	43.48	60 [c]	>0.90
	Rate of movement	Wilcoxon	2.34	8 [b]	>0.95
	Core home range size	Wilcoxon	0.84	8 [b]	>0.999
	Fission/fusion	Wilcoxon	1.15	8 [b]	>0.995

[a]A total of 35 individual bulls were analysed across all four hunts, giving a degrees of freedom value of number of tests = 70.
[b]Bulls and breeding herds were analysed collectively for the four hunts, giving a degrees of freedom value of number of tests = 8.
[c]A total of 30 independent breeding herds were analysed across all four hunts, giving a degrees of freedom value of number of tests = 60

For individuals not associated with the hunted animal, average faecal stress hormone metabolite concentrations increased significantly relative to their respective baseline values in both the four-day (Wilcoxon signed-ranks test: T = –2.29, N = 10, P = 0.022) and the one-month (Wilcoxon signed-ranks test: T = –0.29, N = 10, P = 0.026) periods following the four hunt events (Figure 2) for the data pooled (averaged) for each of the individuals involved in more than one hunt. Individually, 11 of 14 bulls showed an increase in their average faecal stress hormone metabolite concentrations in the four-day period following the hunt events relative to their baseline faecal stress hormone metabolite levels, and 12 of these bulls experienced increased average levels of faecal stress hormone metabolites relative to their baseline stress levels in the one-month period following the four hunt events (Figure 2).

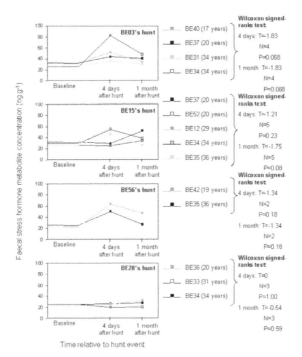

Figure 2. The effect of bull hunts on bulls not present at the hunt. We present the faecal stress hormone metabolite concentrations of individual bulls not associating with the targeted bulls at the time of their hunts. There was no significant increase between baseline: four day and baseline: one month average stress levels of individuals (Wilcoxon signed-ranks test: P>0.05 for all). When individuals were combined, there was a significant increase in baseline: four day and baseline: one month average stress levels (Wilcoxon signed-ranks test: P<0.05 for both).

The individual bulls' faecal stress hormone metabolite concentrations were higher during the four-day post-hunt period as opposed to the one-month post-hunt period for BE03's, BE15's and BE56's hunts (Figure 2). This can be explained in terms of the fact that there were other bulls associating with these targeted individuals at the time of them being hunted. It is likely that these non-hunted bulls emitted distress vocalisations for some period of time following the hunt event. Since elephant vocalisations can be transmitted over relatively long distances (e.g. [13], [14]), it is possible that these bulls' distress calls were received by the remaining bull population, resulting in a general increase in stress being induced. There were no real effects on other bulls from BE28's hunt (Figure 2).

All of the bulls for which faeces were collected experienced elevated stress hormone concentrations in response to the four hunting events. This is in contrast to the behavioral observations, where only 11% of the bulls observed were found to significantly change their movement dynamics in terms of their distances from the respective hunt sites in the ten-day period following the hunt events, and only half of them increased their core home range sizes in the one-month period following the hunt events. Thus behavioral observations alone did not comprehensively quantify the effects of hunting individuals on the remaining bull population as behavioral responses relative to the hunt events were only observed for some individuals whereas increased physiological stress was detected for all individuals.

During the 'before' hunt periods, the adult females had higher frequencies of 'low' stress levels while in the period following the hunt adult females had higher frequencies of 'intermediate' stress levels (Figure 3). This indicated a general physiological stress response to the hunts. However, analyses of the breeding herds' movement dynamics showed no significant changes in response to the hunt events.

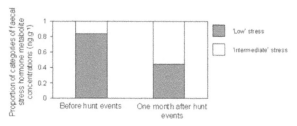

Figure 3. The effect of bull hunts on breeding herds. We present stress levels 'before' and 'one month after' the bull hunts as adult females' categorised faecal stress hormone metabolite concentrations ('low' = 6.3–40.67 ng.g-1 and 'intermediate' = 40.68–75.05 ng.g-1; no 'high' concentrations were recorded).

We observed twelve herds before and after each hunt, and assessed a fission/fusion response by measuring the percentage of time that they spent with other herds (fusion) or alone (fission). Breeding herds were found significantly more on their own as opposed to in groups of herds in the ten-day periods following the hunt events (Wilcoxon signed-ranks test: P<0.05 for all) (Table 4).

Table 4. The effect of the four hunts on the independent breeding herds' fission and fusion dynamics in the ten-day period before and after each hunt event.

ID	Hunt							
	BE03		**BE15**		**BE56**		**BE28**	
	Before (%)	After (%)	Before (%)	After (%)	Before (%)	After (%)	Before (%)	After (%)
CE01	0	0	100	80	0	100	100	100
CE02	0	25	0	0	0	0	0	7
CE03	0	0	40	80	0	0	12	17
CE07	0	33	0	75	12	0	0	2
CE17	0	0	0	0	24	100	67	11
CE19	0	50	0	0	0	0	0	0
CE20	20	60	0	0	33	44	0	0
CE32	0	0	100	100	100	25	100	33
CE54	0	0	0	0	0	8	56	0
CE56	0	0	0	0	6	14	67	17
CE57	0	0	0	0	32	0	40	0
CE59	0	0	0	0	40	33	0	75

Data are percent that each herd was seen with other herds out of the total number of sightings for that herd in the given time period (larger values indicate greater association, 0 indicates that the herd was seen, but was always on its own).

Discussion

In response to short duration hunting events, wildlife often exhibit short-term behavioral responses; however, corresponding physiological responses have not been measured. Behavioral responses by wildlife to hunting activity includes directed movement away from hunters [15], changes in diet [16], distributional shifts [17], greater use of vegetative cover [15] or shifts in core area use within the home range [18]. Behavioral responses to hunting might also be related to prior experiences. For example, elk (Cervus elaphus) became less tolerant of hunter activity later in the hunting season, resulting in more extreme movements as the season progressed [15]. Despite a consistent behavioral response, we are not aware of any study documenting a physiological stress response to hunting. For example, in Spain, Dalmau et al. [19] did not find a relationship between hunting activity and stress in Pyrenean chamois (Rupicapra pyrenaica pyrenaica), as measured by faecal glucocorticoid metabolites. Despite the behavioral responses noted above, Millspaugh et al. [8] did not observe a correlation between hunting activity and faecal glucocorticoid metabolites in elk. Extending hunting hours did not increase corticosterone levels in mourning doves (Zenaida macroura) when compared with baseline values [20]. These studies of physiological stress collectively point to the importance of physical and environmental (e.g., weather) stressors. In contrast, we hypothesize that stress in elephants, which have a complex social system, might be heavily influenced by psychosocial stressors that result from hunting activity. Social vertebrates often exhibit increases in stress due to psychosocial stressors [21].

Many private reserves and protected areas in southern and eastern Africa make use of trophy hunting for income generation, and as a means of both eliminating 'problem' elephant bulls [22], [23] and manipulating the population's linear dominance hierarchy [24]. This is the first detailed study quantifying the effects of bull hunting on the remaining elephants. One might expect, particularly in a relatively small, confined population, that such direct management intervention would significantly affect the remaining elephants. However, only one major reaction occurred, as a result of a poor kill, and the only notable minor behavioral effects were short-term (one day) 'flight' responses by bulls present at the hunt. Physiologically, the induced stress response (both short and longer-term) manifested throughout the population, indicating communication amongst individuals, and may represent the reaction of elephants to each other's suffering (see [25]). The response from those individuals that were present at the hunts of the targeted bulls may have been transmitted by means of a 'domino-effect' throughout the remaining bull population. Interestingly, our results indicate transmission of stress events from bulls to cows.

Limited behavioral studies investigating the specific effects of direct human impact on elephants have been conducted. Poaching pressure leads to increasing group size [26], and 'migration' and aggregation into protected areas [27]. Culling causes disturbance [28], with elephant from a culled population shifting to drink more at night [22]. Four out of the 10 collared female elephants that were within 7 km of a culled group undertook extreme direct movements of 23, 25, and 30 km overnight or within two days of the cull, and which took them out of their then pre-determined range [29].

The results of this study can be applied by elephant managers in small reserves, and may contribute to future debates regarding the implementation of elephant population control through selective bull hunting (see [30] for a treatment of the broader ethical question of hunting elephants). We conclude that hunting bulls in the manner and frequency described here will have no major effect on the remaining elephant population. Despite the increase in faecal glucocorticoid metabolites, the stress response in elephants was short-lived and, in our opinion, not detrimental (the peaks being lower than that shown to natural extreme stressors such as transport or extreme, loud, noises including thunderstorms [9]). However, bulls should preferably not be hunted when they are in musth as they are more aggressive and unpredictable than usual [31], [32]. Because of observed stress responses of bulls present at the hunt, best practice to reduce unnecessary stress indicates that bulls should only be hunted when they are alone. We also recommend that sufficient time (in this study found to be one month) between direct disturbance events be allowed to reduce the possibility of chronic stress (i.e. cumulative effects which were not assessed in this study) in the population. However, we also caution that the effects of disruption of the dominance hierarchy, through hunting, on stress levels, movements and other behavior has not been investigated, and may, in sustained hunting situations, be significant.

Materials and Methods

Pilanesberg National Park (25°8'S–25°22'S; 26°57'E–27°13'E, 560 km2) is located in the North West Province of South Africa. The area comprises predominantly hilly terrain and falls within the transition zone between the Kalahari Thornveld in the west and the Bushveld in the east. The habitat comprises Acacia and broadleaf bushveld, which ranges from closed thickets to open grasslands. The general vegetation type is classified as sourveld [33]. One major river system runs through the centre of the Park, and numerous non-perennial tributaries and streams and several small dams are scattered throughout. Rainfall occurs in summer, and is approximately 630 mm per annum. Temperatures range from a minimum of 1–5°C in winter to a mean maximum of 28–31°C in summer. The park border fence is electrified and provides an effective barrier to elephants (no breakouts to date).

Elephants were introduced, mainly from Kruger National Park between 1981 and 1998 [34]. In 1998, when the population was comprehensively identified to the individual level for the first time, there were 93 elephant, including 17 individually recognizable independent males. These were males that had left the female groups, and were consistently alone or with other males. These ranged in age from 18 to 25 years old. There were also six older males (up to 35 years old) introduced to solve the rhino-killing problem (see [34]). By March 2002, there were 163 elephant, including 39 adult bulls between 12 and 40 years old, and 12 known breeding herds.

Pilanesberg National Park leases out a hunting concession on an annual basis. The Board and Park management are responsible for formulating a hunting quota per species, which is revised annually based on species' abundance and the Park's objectives. Two elephant hunts are usually sold per annum, where each hunt generates an average of $10 000 (approximately R70 000). This revenue is fed back into the Park management.

In 1996, elephant bulls were identified as a major source of both black and white rhinoceros mortality in the Park [24]. The Park controlled these 'problem animals' by hunting two known culprits [34], as well as five problem animals (chasing rhino or damaging vehicles or infrastructure) between 1996 and 2001.

A total of eleven bulls were hunted from 1996 through 2003, with the last four hunts being intensively studied. (1) BE03 (prefix 'BE' denotes bull and 'CE' cow elephants, with a unique individual numerical code following) was hunted on the 16 April 2002. BE15 and BE30 were with him at the time of the hunt. (2) BE15 was hunted on the 30 July 2002. BE05 was with him at the time of the hunt. BE43 and CE01's herd were within 2 km of the hunt. (3) BE56 was hunted on the 9 May 2003. BE05, BE29 and BE37 were with him at the time of the hunt. (4) BE28 was hunted on the 7 August 2003. He was alone at the time of the hunt. A professional hunter, a Park representative, and a ground crew of at least four individuals always accompanied the client, and hunts took place from early to mid-morning. In ten of the 11 hunts, the bull was killed cleanly and went down almost immediately. The exception was a musth bull hunted in 1996, who was wounded by the client and killed by the professional hunter. For the four intensive hunts, targeted bulls were followed intensively in the week leading up to a hunt, and we recorded the bull's locations, associations with other bulls and breeding herds, and daily displacement rate.

Behavioral Observations

The effects of the hunts on the elephant population were divided into 'major' and 'minor' events: 'Major' events included elephants breaking out of the Park, causing damage to infrastructure, and being responsible for tourist-related incidents (e.g. damaging vehicles, increased aggression etc.) associated with the hunt events. 'Minor' events referred to less obvious responses of the remaining elephants to the hunt events. Since movement is often the easiest parameter to measure when assessing the responses of animals to disturbance, we recorded the elephants' movement dynamics in response to the hunt events. The effect of the hunt on the movement dynamics of the remaining elephant population was divided into the short- and longer-term, with the short-term referring to the ten-day period before and after the hunt and the longer-term being the period one month before and one month after the hunt. Only those bulls and individual breeding herds that were located at least ten times each in the month before and after the hunt were used in the analysis. No unusually different behavior was noted in any of the other animals that were excluded from the analysis for lack of sufficient sample sizes.

All elephants were identified using distinctive markings. Each of the matriarchs of the twelve herds were radio-collared with VHF transmitters, which facilitated locating them. A concerted effort was made to locate as many individual elephants as possible each day, and particularly for uncollared males, tourist guides criss-crossing the reserve (and who were familiar with the identities of individual males) helped with locations. All locations were confirmed by us. We searched intensively to find individuals that were with the hunted animal, or close-by. Each individual bull's and breeding herd's locations for the period ten days before and ten days after the hunt were mapped using Animal Movement extension in ArcView 3.1 (ESRI) [35], and a polyline indicating the direction moved between each point was created. The distance between each location point and the hunt site was calculated in ArcView and converted to km. Distances before and after each hunt event were contrasted using the Mann-Whitney U-test, with each individual representing independent data points.

The direction of movement relative to the hunt site was calculated manually using printouts of each individual's movements from ArcView. The direction moved between two points (see Figure 1 x-axis for time interval between points for each individual per hunt) was determined by using the straight line between the first point and the hunt site as a reference line (i.e. 0° line on the protractor) and determining the angle between the reference line and the straight line drawn between the first and second points. This was done on a 0–180° scale, with 0–90° indicating movement towards the hunt, and 90–180° indicating movement away from the hunt. Each value was then assigned as 'positive' (towards the hunt) or 'negative' (away from the hunt). Changes in direction moved before and after hunts were assessed using sign tests, with the average each individual representing independent (paired) data points. The direction data were then plotted against the distance data to illustrate the direct (short-term) effect of the hunt on the movement dynamics of each individual bull or breeding herd.

The displacement rate of each bull and breeding herd in the ten-day period before and after each hunt event was determined by calculating the distance between each successive location point in ArcView, and then dividing the distance by the respective time interval. To remove any bias in the analysis for those individuals where more than one location per day was obtained, the first datum recorded per day was used in the calculation of rate of displacement. For the analysis of those bulls that were present at the actual hunting of the targeted individual, their first recorded displacement was their departure from the hunt site. Since only the first location for these bulls was used, the results would only be biased conservatively in favour of them increasing their distance from the respective hunt sites. Because the times of locations for each individual were different, the displacements were calculated and expressed in terms of $km.h^{-1}$. The coefficient of variation (CV) [36] was calculated for each individual bull and breeding herd's displacement rate before and after the respective hunt events, with a higher CV indicating more erratic displacement rates. Displacement rates before and after each hunt event were contrasted using the Mann-Whitney U-test, with average values for each individual representing independent data points.

To determine whether the breeding herds showed a 'fission' or 'fusion' (i.e. whether the herds came together or dispersed) response following the hunts, the number of matriarchs (where one matriarch indicates the presence of one herd) seen together in the ten-day period before the respective hunt and in the ten-day period after the hunt was compared. We calculated the total number of sightings of each herd in the ten-day period before and after the hunt, and dividing these sightings into whether the herd was alone or with other herds. The frequency of being with other herds was calculated, so that each herd had one value representing their 'grouping' tendency before and one value representing their 'grouping' tendency after the hunt. Thus each independent datum represented a single breeding herd. The percentage sightings with other herds (fusion) was contrasted before and after each hunt event using the Wilcoxon test, with each herd representing independent (paired) data points.

Each individual bull's and breeding herd's locations (using the first location per day) for the period of one month before and one month after the hunt were mapped in ArcView (ESRI). Separate fixed kernel home ranges [35], using the 95 % and 50 % probability contours, were plotted for the month before and after the hunt. These were then overlayed, and the percentage area overlay from the 'before hunt' home range and the 'after hunt' home range was computed to determine whether a shift in home range (long-term effect) had occurred subsequent to the hunt. The core home range (area enclosed by the 50 % probability contour) from before and after the hunt were also determined, and represented as a factor increasing or decreasing relative to the 'before hunt' core home range area. Range sizes were contrasted before and after each hunt event using the Wilcoxon test, with each individual representing independent (paired) data points.

Data were not normally distributed (Kolmogorov-Smirnov test: P<0.05), and thus non-parametric tests (Sign test, Mann-Whitney U test, Wilcoxon signed-ranks test) were used in the statistical analyses.

The overall patterns of behavioral response was tested using the Combined Probabilities Test. The 'combined probabilities' analysis allows for separate significance tests on different data sets that test the same scientific hypothesis to be combined and analysed collectively [37]. The probability values obtained from the four separate analyses conducted for each of the individual bulls and the five analysis for each of the independent breeding herds relative to the hunt events (i.e. distance from hunt site, direction of movement relative to hunt site, displacement rate, core home range sizes and fission and fusion responses (breeding herds only)) (see above for detail) were tested using combined probabilities [37]. These analyses reflect the more robust assessments of behavioral responses as the problem of pseudoreplication, which may be present in the individual based analyses, is largely negated.

Physiological Stress Response

Dung samples were collected throughout the study period from March 2002 to July 2003. The protocol for collecting, storing and processing the samples and the extraction of cortisol from the samples is extensively described elsewhere [11], [38].

Upon collection, each sample was allocated a unique numerical code and the date, time of sample collection, bolus measurements (top diameter, bottom diameter and length), location (GPS co-ordinates and name of road nearest to sample), and the identification of the elephant known to have defecated were recorded. Some samples were collected without knowing which individual had deposited them. These samples were 'sexed' and aged by (1) referring to the spoor around the sample and (2) using the diameter of the bolus where adult (>15 yrs) bulls' boluses were generally found to have a minimum diameter of 12 cm, while adult (>15 yrs) cows' boluses generally had a minimum diameter of 10 cm, sub-adults (6–15 years) a minimum diameter of 6–9 cm, and juveniles (less than 6 years) a minimum diameter of less than 6 cm , based on samples from known individuals (pers. obs.).

The actual time of sample collection was corrected per sample based on the estimated age of the sample. Thus if the sample was estimated as being 5 h old, then 5 h were subtracted from the 'collected' time to give a 'corrected' time.

A 'lag-time' of 36 h (the time taken for a 'stressful' event to be maximally detected in African elephant faeces [7], was used to correlate specific events with the faecal stress hormone metabolites present in the dung samples. We subtracted 36 h from the 'corrected time' of each sample to give the actual time at which the stress occurred.

Samples were categorised into 'low', 'intermediate' and 'high' stress by determining the total range of faecal stress hormone metabolite concentrations from all samples collected (6.3–109.43 ng.g^{-1}), the physiological stress response of elephants [7], [9], and dividing them into three parts. Thus samples with 6.3–40..67ng.g^{-1} stress hormone metabolite concentrations = 'low' stress, samples with 40.68–75.05ng.g^{-1} stress hormone metabolite concentrations = 'intermediate' stress, and samples with 75.06–109.43ng.g^{-1} stress hormone metabolite concentrations = 'high' stress. The basis for these categorisations is supported by the ACTH challenge results [7], [9].

There were a total of six bulls that were associating with the respective hunted bulls (within 10 m of the targeted bull) at the time of the hunt events. Baseline faecal stress hormone metabolite concentrations were obtained for each of these six bulls by averaging their respective 'low' stress level faecal metabolite concentrations from one month before the hunt event (only the 'low' stress level category was used in order to account for individual variation in the stress response (e.g. [39], [40], [41]). The time taken for each of these six bulls to return to their respective baseline faecal stress hormone metabolite concentrations was determined by plotting their actual stress level concentrations against time (in terms of days after hunt event). The baseline faecal stress hormone metabolite concentrations for these six bulls after the hunt events was calculated by averaging their 'low' stress level faecal metabolite concentrations for one month after their metabolite concentrations had returned to their former baseline levels. The baseline faecal stress hormone metabolite concentrations of the individual bulls in the one-month before the hunt were tested for significant differences with (i) their baseline faecal stress hormone metabolite concentrations in over one-month after the hunt (i.e. controlling for change over time), and (ii) with the average of the maximum three faecal stress hormone metabolite concentration values obtained during the time that it took for each of the individuals to return to their baseline faecal stress hormone metabolite concentrations after the hunt. The Wilcoxon signed-ranks test was used.

Sufficient faecal samples in the one-month period following each hunt event were collected from fourteen individually identified adult bulls from those not associating with the targeted bull at the time of his hunt. Four of these bulls were 'replicates' in that sufficient samples were collected from each of them relative to more than one hunt. The Analyses were conducted for a four-day (short term) and a 5–30 day (long-term) period following each hunt event. The average of the three maximum values obtained for each individual was used. Baseline faecal stress hormone metabolite concentrations for each individual were calculated using the remaining samples in the 'low' stress level category collected throughout the broader study period. The 4 and 5–30 day post-hunt values were tested for significance against the baseline average faecal stress hormone metabolite concentration value using the Wilcoxon signed-ranks test with each individual (paired) representing independent data points.

Insufficient identified faecal samples from individuals within a particular breeding herd precluded using an individual-based analysis for female responses. Instead, the adult (15 years and older, to exclude the possibility of age confounding the results) breeding herd animals' faecal stress hormone samples were combined to give a general breeding herd response to hunt events. This is made possible due to the fact that (1) members of the same herd show synchronous increases in stress hormone metabolites in response to 'stressful' events [42, Millspaugh, unpublished data], and (2) the strong social bonds existing between different family groups (e.g. [6]) and the fact that elephants can communicate over relatively long distances (e.g. [13], [14]) allow for the assumption that this synchronicity in stress hormone production will extend throughout the breeding herd population, particularly in a relatively small reserve such as Pilanesberg National Park. Data for one month following each of the four hunts were categorised as being 'After' the hunt events. The remaining data were classified as being representative of 'Before' the hunt events. The respective samples were classified into 'low', 'intermediate' and 'high' levels of stress. We used log-linear analyses [43] of the different categories ('Low', 'Intermediate' and 'High' stress) before versus after the hunt.

Acknowledgements

We thank all of the people who assisted with locating elephants during the study, and particularly the staff of North West Parks and Tourism Board for facilitating and contributing to the success of this study. We thank those who assisted with laboratory analysis at Missouri. This study was approved by the Animal Ethics Committee of the University of KwaZulu-Natal.

Authors' Contributions

Conceived and designed the experiments: RS TB GV. Performed the experiments: TB GV. Analyzed the data: RS TB. Contributed reagents/materials/analysis tools: JM. Wrote the paper: RS TB BP JM. Other: Performed the Hormone Assays: JM.

References

1. Peterson MN (2004) An approach for demonstrating the social legitimacy of hunting. Wildlife Society Bulletin 32: 310–321.

2. Lecocq Y (1997) Elephant culling. Wildlife Society Bulletin 25: 215.

3. Reynolds P, Braithwaite D (2001) Towards a conceptual framework for wildlife tourism. Tourism Management 22: 31–42.

4. Mbenga E (2004) Visitor wildlife viewing preferences and experiences in Madikwe Game Reserve, South Africa. M. Env. Manag. Dissertation, University of KwaZulu Natal, South Africa.

5. Breytenbach G, Sonnekus I (2001) Conservation business: sustaining Africa's future. Koedoe 44: 105–123.

6. Poole JH (1995) Sex differences in the behavior of African elephants. In: Short RV, Balaban E, editors. The differences between the sexes. Cambridge University Press. pp. 331–346.

7. Wasser SK, Hunt KE, Brown JL, Cooper K, Crokett CM, Bechert U, Millspaugh JJ, Larson S, Monfort SL (2000) A generalized fecal glucocorticoid assay for use in a diverse array of nondomestic mammalian and avian species. General and Comparative Endocrinology 120: 260–275.

8. Millspaugh JJ, Woods RJ, Hunt K, Raedeke KJ, Washburn BE, Brundige GC, Wasser SK (2001) Using fecal glucocorticoids to quantify the physiological stress response of free-ranging elk. Wildlife Society Bulletin 29: 899–907.

9. Millspaugh JJ, Burke T, van Dyk G, Slotow R, Washburn BE, Woods R (2007) Stress response of working African elephants to transportation and Safari Adventures. Journal of Wildlife Management 71: 1257–1260.

10. Harper JM, Austad SN (2000) Fecal glucocorticoids: a non-invasive method of measuring adrenal activity in wild and captive rodents. Physiological and Biochemical Zoology 73: 12–22.

11. Millspaugh JJ, Washburn BE (2004) Use of fecal glucocorticoid metabolite measures in conservation biology research: considerations for application and interpretation. General and Comparative Endocrinology 138: 189–199.

12. Creel SC, Fox JE, Hardy A, Sands J, Garrott RA, Peterson RO (2002) Snowmobile activity and glucocorticoid stress responses in wolves and elk. Conservation Biology 16: 1–7.

13. Langbauer WR Jr, Payne KB, Charif RA, Rapaport L, Osborn F (1991) African elephants respond to distant playbacks of low-frequency conspecific calls. Journal of Experimental Biology 157: 35–46.

14. McComb K, Moss CJ, Sayialel S, Baker L (2000) Unusually extensive networks of vocal recognition in African elephants. Animal Behavior 59: 1103–1109.

15. Millspaugh JJ, Raedeke KJ, Brundige GC, Gitzen RA (2000) Elk and hunter space use sharing in the Southern Black Hills, South Dakota. Journal of Wildlife Management 64: 994–1003.

16. Morgantini LE, Hudson RJ (1985) Changes in diets of wapiti during a hunting season. Journal of Range Management 38: 77–79.

17. Irwin L, Peek JM (1979) Relationships between road closures and elk behavior in northern Idaho. In: North American elk: ecology, behavior and management. Boyce MS, Hayden-Wing LD, editors. University of Wyoming, Laramie, Wyoming, USA. pp. 199–204.

18. Kilpatrick H, Lima KK (1999) Effects of archery hunting on movement and activity of female white-tailed deer in an urban landscape. Wildlife Society Bulletin, 27(2): 433–440.

19. Dalmau A, Ferret A, Chacon G, Manteca X (2007) Seasonal changes in fecal cortisol metabolites in Pyrenean chamois. Journal of Wildlife Management 71: 190–194.

20. Roy C, Woolf A (2001) Effects of hunting and hunting-hour extension on mourning dove foraging and physiology. Journal of Wildlife Management 65: 808–815.

21. Sapolsky R (1987) Stress, social status, and reproductive physiology in free-living baboons. In: Crews D, editor. Psychobiology of reproductive behavior: an evolutionary perspective. Prentice Hall, Englewood Cliffs, New Jersey, USA. pp. 291–392.

22. Martin RB, Craig GC, Boot VR (1996) Elephant management in Zimbabwe. Department of National Parks and Wildlife Management. Harare, Zimbabwe.

23. Boonzaaier WW, Collinson RFH (2000) Pilanesberg National Park management plan. Second edition. Pilanesberg National Park Management Series, Contour Project Managers and Collinson Consulting, Rustenburg, South Africa.

24. Slotow R, van Dyk G (2001) Role of delinquent young 'orphan' male elephants in high mortality of white rhinoceros in Pilanesberg National Park, South Africa. Koedoe 44: 85–94.

25. Douglas-Hamilton I, Bhalla S, Wittemyer G, Vollrath F (2006) Behavioral reactions of elephants towards a dying and deceased matriarch. Applied Animal Behavior Science 100: 87–102.

26. Eltringham SK (1977) The numbers and distribution of elephant Loxodonta africana in the Rwenzori National Park and Chambura Game Reserve, Uganda. East African Wildlife Journal 15: 19–39.

27. Western D (1989) The ecological role of elephants in Africa. Pachyderm 12: 42–45.

28. Cumming D, Jones B (2005) Elephants in southern Africa: Management issues and options. WWF-SARPO Occasional Paper Number 11, Harare.

29. Whyte I (1993) The movement patterns of elephants in the Kruger National Park in response to culling and environmental stimuli. Pachyderm 16: 72–80.

30. Lötter HPP, Henley M, Fakir S, Pickover M, Ramose M (in press) Ethical considerations in elephant management. In: Scholes RJ, Mennell KG, editors. Assessment of South African Elephant Management. Johannesburg: Witwatersrand University Press. pp. 307–338.

31. Hall-Martin AJ (1987) Role of musth in the reproductive strategy of the African elephant (Loxodonta africana). South African Journal of Science 83: 616–620.

32. Poole JH (1989) Mate guarding, reproductive success and female choice in African elephants. Animal Behavior 37: 842–849.

33. Acocks JPH (1988) Veld types of South Africa. Memoirs of the Botanical Survey of South Africa. Third edition. Botanical Research Institute, Department of Agriculture and Water Supply, South Africa.

34. Slotow R, van Dyk G, Poole J, Page B, Klocke A (2000) Older bull elephants control young males. Nature 408: 425–426.

35. Hooge PN, Eichenlaub B (1997) Animal movement extension to ArcView version 1.1. Alaska Biological Science Center. Anchorage: U.S. Geological Survey.

36. Zar JH (1999) Biostatistical analysis. Fourth edition. Prentice Hall, New Jersey, USA.

37. Sokal RR, Rohlf FJ (1981) Biometry. Second edition. W.H. Freeman, New York.

38. Millspaugh JJ, Washburn BE, Milanick MA, Slotow R, van Dyk G (2003) Effects of heat and chemical treatments on fecal glucocorticoid measurements: implications for sample transport. Wildlife Society Bulletin 31: 399–406.

39. Brown IRF White PT (1979) Serum electrolytes, lipids and cortisol in the African elephant (Loxodonta africana). Comparative Biochemistry and Physiology 62: 899–901.

40. De Villiers MS, Meltzer DGA, van Heerden J, Mills MGL, Richardson PRK, van Jaarsveld AS (1995) Handling-induced stress and mortalities in African wild dogs (Lycaon pictus). Proceedings of the Royal Society of London Bulletin 262: 215–220.

41. Moberg GP, Mench JA (2000) The biology of animal stress. Basic principles and implications for animal welfare. CABI Publishing, New York.

42. Pretorius Y (2004) Stress in the African elephant on Mabula Game Reserve, South Africa. M.Sc. Thesis, University of KwaZulu-Natal, Durban, South Africa.

43. Knoke D, Burke PJ (1980) Log-linear models. In: Lewis-Black MS, editor. Series: Quantitative applications in the social sciences. Sage Publications Incorporated, California, USA. pp. 9–79.

CITATION

Home Range Utilisation and Territorial Behavior of Lions (Panthera leo) on Karongwe Game Reserve, South Africa

Monika B. Lehmann, Paul J. Funston, Cailey R. Owen
and Rob Slotow

ABSTRACT

Interventionist conservation management of territorial large carnivores has increased in recent years, especially in South Africa. Understanding of spatial ecology is an important component of predator conservation and management. Spatial patterns are influenced by many, often interacting, factors making elucidation of key drivers difficult. We had the opportunity to study a simplified system, a single pride of lions (Panthera leo) after reintroduction onto the 85 km2 Karongwe Game Reserve, from 1999–2005, using radio-telemetry. In 2002 one male was removed from the paired coalition which had been present for the first three years. A second pride and male were in a fenced reserve adjacent of them to the east. This made it possible to separate social and resource factors in both a coalition and single male scenario, and the driving factors these seem to have on spatial ecology. Male ranging behavior was not affected by coalition size, being driven more by resource rather than social factors. The females responded to the lions on the adjacent reserve by avoiding the area closest to them, therefore females may be more driven by social factors. Home range size and the resource response to water are important factors to consider when reintroducing lions to a small reserve, and it is hoped that these findings lead to other similar studies which will contribute to sound decisions regarding the management of lions on small reserves.

Introduction

Interventionist conservation management of territorial large carnivores has increased in recent years, especially in South Africa, where farmland has been rehabilitated to game reserves and many species have been reintroduced [1]–[3]. At least 37 reserves have reintroduced lions (Panthera leo) primarily for ecotourism, but also for ecological processes [3]–[5]. Understanding of spatial ecology is an important component of these two management objectives, both in planning prior to reintroduction, and in subsequent population management to ensure that the population introduced is not above carrying capacity and is representative of a natural population in terms of size and structure. Because these reserves manipulate both the resource (e.g. water provision and harvesting of prey species) and social environment (selective removals or supplementation of lion), separating the different competing drivers of spatial ecology is important in order to make the correct management decisions in such small reserves.

The home range of a carnivore is generally as large as is necessary but as small as possible to satisfy energetic needs [6], [7]. Upper limits are determined by energy expenditure during territorial defence [8] while lower limits are governed by food availability [9]. Adult male lions maintain a territory largely contiguous with that of their home range and discourage rivals from entering these by patrolling, scent-marking and roaring [8]–[10]. Territorial males can protect their cubs from infanticide [11] in two ways: either directly by accompanying the pride and chasing out rival males [10], [12], or indirectly by maintaining the security of a territory [10].

Male lions show territorial behavior by roaring and scent-marking while patrolling. Territorial displays are expensive because they separate the males from their females, increasing the risk of infanticide by invading males [13], [14].

Furthermore, roaring highlights the location of the males for intruding coalitions [15], [16], and is also energetically expensive, both in terms of the distances covered [8], and the energetic cost of roaring [17].

The two factors, advertisement and resources may influence territory size, shape, and usage in different ways [18], [19]. It is extremely difficult to separate out these two potentially confounding factors in natural circumstances. However, we had the opportunity to test their relative influences in an artificial situation where a single pride male and pair of females existed in a fenced reserve, and a second pride and male were in a fenced reserve adjacent of them. The fences were electrified and effectively lion-proof. Although from a small sample set, we were therefore able to separate our predictions of responses to social influences and resources, and assess their relative input to lion behavioral decisions and subsequent costs.

We predicted that lions would respond towards resources such as prey, water and cover in all directions, but would respond to social influences only in the direction of the adjacent pride. We measured male and female range use, as well as male scent-marking and roaring in different parts of their range. We assess the relative influence that the social factors impose on the resource factors. We had the further advantage of an experimental manipulation of the system, whereby the social system was manipulated by the removal of one of the coalition males while holding resources constant.

Methods

Study Area

Fieldwork was conducted on the 85 km² Karongwe Game Reserve (24°13'S and 30°36'E), located in the Limpopo Province, South Africa (Fig. 1). Altitude here varies from 489 m to 520 m above sea level. The reserve falls within the Savanna Biome [20] and lies within the Mixed Lowveld Bushveld [21]. The Greater Makalali Conservancy borders Karongwe on its eastern boundary and is the only other reserve in the area that supports lions, which were introduced to Makalali in 1994 (Fig. 1). Both reserves are fenced with lion-proof electric fences, and there is a road (15 m wide area) separating the two adjacent fences.

As the reserve's main function is eco-tourism a large number of species are present. Apart from lions, other large carnivores include leopards (Panthera pardus), cheetahs (Acinonyx jubatus), wild dogs (Lycaon pictus) and spotted hyeanas (Crocuta crocuta), and there are twelve ungulate species as potential prey.

Figure 1. Location of Karongwe Game Reserve and the neighbouring Makalali Game Reserve in Limpopo Province, South Africa.

Field Data Collection

The study was conducted over a six year period from October 1999 to October 2005, totalling 2192 field days. During that time the lion population varied from four to a maximum of 11 individuals, with an average of eight. After three years, one of the males was removed from the coalition as they were considered to be removing too much prey [22].

A member of each subgroup was located twice daily where possible for the duration of the study (92.3%, n = 2024 days), using the standard method for radio telemetry tracking [23]. Most observations took place between 5:00–10:30 (48.4%, n = 1809) and 15:30–20:30 (47.7%, n = 1782), with some observations at night (3.9%, n = 147). The nocturnal observations only covered the period 2004 to 2005 and took place from 22h00 to 02h00. A specific nocturnal study was undertaken to focus on recording territorial behavior by the pride male from February 2005 to June 2005 (n = 48 nights, 12 sessions of four consecutive nights). Three shift times were chosen: 17:00–23:00, 21:00–04:00, and 23:00–06:00 to incorporate dusk, the middle of the night, and dawn. One session for each shift time in each of the four moon phases was completed. Four nights were spent following the lions continuously from 17:00–06:00.

After locating the focal animal, the following data were recorded: date, time, location, GPS co-ordinate, daily belly score, and general behavior. The male was followed by vehicle at a distance of 15–30 m. Lions were viewed using a spotlight with a red filter. Any territorial behavior activities were recorded and georeferenced. All movement and most territorial data were unfortunately only collected after the removal of one male. Therefore no direct comparisons between the one male and a two male coalition's behavior were possible. We realise that the data set is small (n = 1 pride), but we feel that the study nevertheless provides value as more and more managers are stocking such small reserves in a similar manner and can benefit from the experiences observed on Karongwe.

Data Analysis

The data were imported into Arcview 3.2 for home range analyses and the delineation of animal movement paths using the extension Animal Movement [24]. Home ranges were delineated using 95% kernel home ranges for point distributions, and 50% kernels to delineate core areas [25]–[28]. For purposes of this study areas that were defined by the 95% kernel that fell outside the reserve boundary were clipped as these could not contribute to the home range.

Social versus resource drivers of territorial behavior were contrasted using preference values for behaviors in a 1 km×1 km grid, i.e. we determined frequency of roaring and scent marking in each grid cell, and preference values for each grid cell. Grid preference was calculated as the ratio of use to availability [19], where a value of >1 indicated a preference and <1 indicated avoidance of an area. A two kilometre buffer zone along the boundary with the neighbouring pride was used to differentiate the response to social rather than resource stimuli. Resource limitation and preference was measured on one scale by contrasting observed ranges with available ranges, and also assessing the influence of rivers [29]. The latter was done by determining the preference for areas within 500 m of drainage lines or rivers.

Because all of the observations made in this study are based on a single male/ coalition, a small female group, and a small subadult male group, we do not have a sample of truly independent samples and the conclusions should be viewed as preliminary, and will hopefully stimulate further work in this area.

Results

Description of Ranging Behavior

Two male lions were released onto the reserve in September 1999, and two lionesses were released a month later. The range available to the lions was 80 km^2.

Table 1 shows the change in home range size during the study period. The pride's and the males' core home ranges were concentrated along the rivers (Fig. 2). The pride's home range (95% range 64.4 km² and 50% core 10.3 km²) was larger than the males' home range (95% range 56.3 km² and 50% core 5.0 km²) with the core double the size of that of the males. There was a noticeable difference between the summer home range (November – April) and the area utilised in winter (May – October). The lions (combined pride and males) utilised almost the whole reserve in summer (95% range 77.4 km² and 50% core 10.6 km²) with a large core encompassing a larger area away from the rivers. The winter home range had a 95% range of 58.9 km² and a 50% core of 6.1 km². Almost 70% of the reserve lies within 500 m of generally permanent water, and 58.2% of the 95% summer home range, and 86.8% of the 50% summer core within 500 m of water; while 99.1% of the winter core range was within 500 m of water.

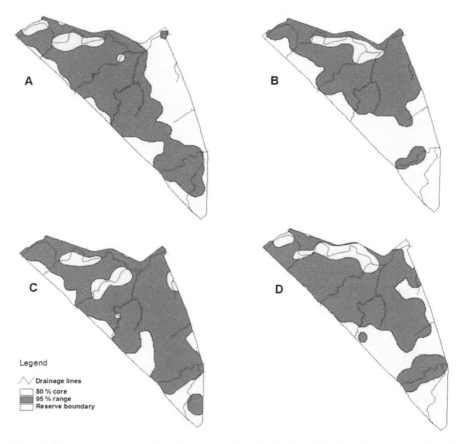

Figure 2. Home ranges of (A) male coalition, (B) single male, (C) females before male was removed, and (D) females after removal of the male.

Table 1. Combined male and female annual minimum convex polygon and kernel home range sizes, illustrating the expansion of the home range from 2000 to 2005.

Home Range	95% range (km²)	50% core (km²)	Minimum convex polygon (km²) *
2000	35.0	1.2	76.9
2001	59.5	5.7	78.1
2002	63.5	7.2	74.7
2003	52.4	4.3	72.9
2004	59.7	5.5	73.6
2005	65.4	5.5	77.9
Management action (splitting male coalition)			
Before	68.8	6.3	78.6
After	66.0	5.1	78.1

*Area of clipped minimum convex polygon that falls within the reserve boundary.

The pride's home range expanded from 53.4 km² to 56.8 km² (Fig. 2). Before the male was removed the lionesses were observed to spend little or no time in the eastern side of the reserve, but included that area thereafter (Fig. 2). The pride males had a home range of 66.6 km² and a core of 6.2 km² that was reduced to 47.4 km² with a core of 4.8 km² when the male was removed (Fig. 2).

Home Range Response to Resource and Social Factors

Despite using an average of 82.2% (69.9–86.6%; n = 4) of the grid cells across the reserve, all lion groups selectively preferred only 21.7–31.3% (n = 4) of these. Almost all the cells in the eastern buffer zone were used (88.9–100%, n = 4) by the four lion groups (Table 2).

Table 2. The percentage of the reserve that is preferred, avoided, or used at availability by the different lion groups measured (a) across the whole reserve to indicate the proportion of the reserve that is utilised, and (b) within the 2 km buffer zone along the eastern boundary to indicate whether the buffer zone is preferred or avoided.

Preference (use/availability)	Male coalition (%)	Single male (%)	Females with coalition (%)	Females with single male (%)
Over the whole reserve				
Preference (>1)	27.7	21.7	26.5	31.3
At Availability (1)	13.3	12.0	10.8	8.4
Avoidance (<1)	59.0	66.3	62.7	60.3
Within the 2 km buffer zone along eastern boundary				
Preference (>1)	38.9	50.0	5.6	5.6
At Availability (1)	16.7	11.1	0	22.2
Avoidance (<1)	44.4	38.9	94.4	72.2

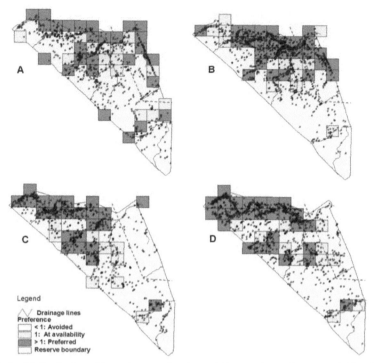

Figure 3. Spatial distribution of lions. Areas preferred, avoided, or used according to availability for (A) male coalition, (B) single male, (C) females before male was removed, and (D) females after removal of the male. The red line indicates the edge of the 2 km buffer zone from the adjacent lion population to the east.

The coalition preferred cells scattered around the reserve and spent more time in those areas than the single male did (Fig. 3). The single male preferred areas mostly in the northern half of the reserve, which coincided with the female's preference, and was also along the major drainage lines.

The areas within 500 m of drainage lines were most preferred by all lion groups with the females showing the highest preference for these areas.

Territorial behavior

Both the coalition and single male showed stronger preference for the buffer area (coalition preferred 38.9% of buffer and only 27.7% of the rest of the reserve; and the single male preferred 50% of the buffer and only 21.7% of the rest of the reserve). Figure 2 indicates that the core ranges were predominantly along drainage lines and indicates preference for those areas, as can also be seen in Figure 3. Point distribution in Figure 3 indicates that both the coalition and single male concentrated on the core areas along the rivers, as well as along the eastern fenceline.

Figure 4 indicates a similar pattern for locations where the males displayed territorial behavior. The males therefore showed the most notable territorial behavior in response to the social factor in the east, as well as resource factors by defending their prime habitat within their home range.

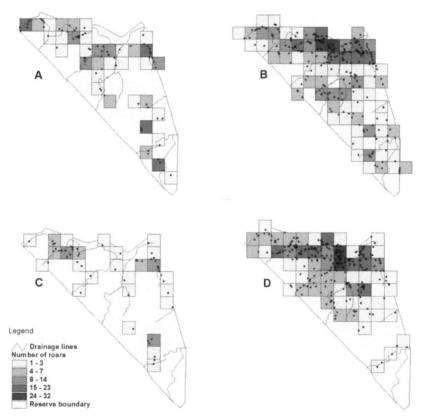

Figure 4. Spatial patterns in territorial behavior indicated by density of behaviors across the reserve. Locations of (A) scent marks made by the pride males from 1999 to 2002, (B) scent marks by the single male from 2002–2005, (C) roars by the pride males from 1999 to 2002 and (D) roars by the single male from 2002–2005.

The coalition preferred, while the single male showed a slightly weaker preference for scent marking within the buffer zone along the eastern boundary. In terms of roaring, both the coalition and the single male strongly preferred using the buffer zone.

Overall 624 scent marks were observed by the pride male/s, as compared with only 119 excretions. Additionally the pride male/s were heard roaring 578 times (Table 3), and 74.3% (n = 1313) of all territorial scent marks and roars occurred within drainage lines or within 500 m of water.

Table 3. The effect of coalition size manipulation (male removal) on territorial behavior of pride males.

Observation period	Days	Scent marks	Roars	Urinate / defecate
Before male removal				
Male	988	126	55	33
Female	988	2	24	20
Total		**128**	**79**	**53**
After male removal				
Male	1130	328	352	58
Female	1130	0	80	28
Subadults	1130	1	14	8
Total		**329**	**446**	**94**
Nocturnal observation after male removal				
Male	48	170	171	28
Female	48	0	62	18
Subadults	48	0	3	1
Total		**170**	**236**	**47**

The single pride male scent marked at a rate of 1.1 scent marks/km. These included spay urinating on bushes (63.2%), of which he rubbed his body or head on the bush before urinating on 34% of the occasions, and urinating on the ground while scraping with the back feet (36.8%).

The single pride male roared at a rate of 0.6 roars/hour (n = 48). The male roared more frequently while alone than with pride members, with an average of 2.0 roars/hour while alone compared to 0.4 roars/hour when not alone.

The pride male covered an average distance of 0.45±0.07 km/hr (n = 48). Figure 5 indicates movement paths on selected nights where more than 4 km were covered. The largest distance covered in one six hour observation was 12.0 km. On nine nights the pride male and any associated lions with him did not walk at all. The lions had a kill on three of those and the male was mating on two of the others. The pride male seemed to cover more distance while on his own than when other members of the pride were with him. This could be a general pattern but equally the small sample size (n = 1) could probably be as good an explanation.

The distances he moved did not seem to be affected by cloud cover, time of night (17:00–00:00; 21:00–04:00; or 23:00–06:00), or the phase of the moon. However, there seemed to be an interaction between time of night and the moon phase, with the furthest distance being walked on full moon between 21:00–04:00 (mean 6.4 km, n = 4), and the shortest distances being walked on new moon between 17:00–23:00 (mean 0.3 km, n = 4). The male walked less than expected between 23:00–06:00 during all four moon phases, and more than expected between 21:00–04:00 during both full moon and the first quarter.

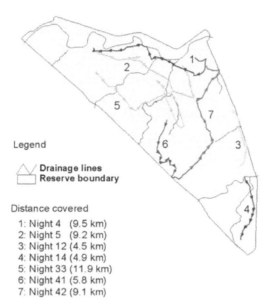

Legend

Drainage lines
Reserve boundary

Distance covered
1: Night 4 (9.5 km)
2: Night 5 (9.2 km)
3: Night 12 (4.5 km)
4: Night 14 (4.9 km)
5: Night 33 (11.9 km)
6: Night 41 (5.8 km)
7: Night 42 (9.1 km)

Figure 5. Selected nights showing movement paths travelled by the pride male, illustrating distance and path walked during territorial patrols.

Discussion

Lion home range sizes vary considerably across study areas, ranging from 20–45 km^2 in places like Manyara National Park and Ngorongoro Crater [30], [31] to 2075 km^2 in arid ecosystems such as Etosha National Park [32]. Even a study on the neighbouring reserve, Makalali, showed variability with home ranges varying from 24.9 km^2 to 106.8 km^2 [33]. Individual variability has been shown to be the largest source of variance in mean estimates of home ranges [34]. On the central and south-eastern basalt regions of Kruger National Park the home ranges were about 100 km2 [17], [35], whereas they were about 250 km^2 on the northern basalt plains [36]. Thus in both Karongwe and Makalali lions seemed to occupy substantially smaller home ranges than in Kruger National Park. This was not due to the fence surrounding the reserve as the lions' overall range sizes were smaller than the reserve potentially allowed.

It has been variously suggested that home range size and configuration of large carnivores is influenced by patterns of resource distribution [7], [37], and by social effects [19]. In this study both the pride males were unaffected by the pride to the east using the buffer area according to its availability, i.e. the males seemed to use the resources of their territory regardless of social influence. The females, however, tended to avoid the buffer area seemingly displaying a negative response to the social influences from the east. Females and their cubs are violently affected by incoming males [38] and could be avoiding the area to prevent contact with potentially infanticidal males. [39] showed that new prides often settle adjacent to their natal range but the pride females on Karongwe, which originated from neighbouring Makalali, were nevertheless cautious of this boundary area. It should be remembered that the study is only based on one pride resulting in a small sample size which could affect the results.

The males and females showed strong preference for drainage lines and rivers. This could be largely due to shifts in prey availability, opportunities for prey capture [29], or cover and protection for cubs [40], particularly as the area was most strongly preferred by the lionesses. Karongwe and Makalali are both small reserves where prey movements are constrained by fences, the only local movements being that prey moves closer to riverine and other water rich areas in winter. It is also important to note that Karongwe was stocked above herbivore carrying capacity. Therefore, there was no resource limitation but the lions still tracked the prey across seasons, indicating that this could be a resource driver of spatial ecology.

Although the pair of males used the buffer zone according to its availability, both the coalition and single pride male showed preference for both scent marking and roaring in this area, reflecting the social influence from the east. The pride male did most territorial patrolling while alone, when he could cover more distance, and not place pride members at unnecessary risk in the event of an encounter. Lionesses were observed to scent mark and roar, but usually while in the presence of the pride male, probably because roaring increases their risk of attracting potentially infanticidal males [38]. [17] also noted that a lot of territorial behavior occurred along drainage lines in the nearby Kruger National Park. This could be because rivers are often natural borders between home ranges [17], or are areas that offer male lions better hunting opportunities [41].

Conclusions

Key results from our study are that male ranging behavior is possibly not affected by coalition size, and suggests it is more driven by resources than social factors. Female ranging on the other hand seemed to be driven by social factors above resources. Males changed their territorial behavior rather than ranging pattern in response to social influences. Although based on a small sample size, these results may have important implications for conservation management of small lion populations. Social effects from surrounding reserves may lead to heterogeneity in ecological influences of lions, and need to be factored into management of prey populations (e.g. harvesting locations). Further, the removal of a male seems to have the advantage of decreasing prey off-take, but does not fundamentally shift ranging behavior, even in the presence of a heterogeneous social influence. This is relevant to shared-ownership reserves that have traversing restrictions for various operators. By the same token, social effects may influence the spatial distribution of tourism-attractive behaviors such as roaring and scent marking. Understanding of fundamental drivers of territorial behavior allows better planning and management of adaptively managed reserves, and these results can potentially contribute to the management of other territorial species.

Acknowledgements

We would like to thank K.e.r.i. Research for the use of unpublished data, and Philip Owen for his hard work coordinating the darting and management of the lions. We thank Sophie Niemann and all K.e.r.i. Research and GVI (Global Vision International) research staff for help with data collection. Thanks go to all Karongwe landowners who allowed the researchers to traverse their land. Dr Peter Rogers is also gratefully acknowledged for his veterinary assistance. Thank you to Vee Cowie for her assistance with GIS.

Authors' Contributions

Conceived and designed the experiments: MBL CRO. Performed the experiments: MBL CRO. Analyzed the data: MBL. Contributed reagents/materials/analysis tools: CRO. Wrote the paper: MBL. Contributed by editing the paper: CRO. Acted as the first author's study supervisor: PJF. Guided the analysis process: PJF RS. Edited the paper: PJF CRO RS.

References

1. Gusset M, Ryan SJ, Hofmeyr M, van Dyk G, Davies-Mostert HT, et al. (2008) Efforts going to the dogs? Evaluating attempts to re-introduce endangered wild dogs in South Africa. J Appl Ecol 45: 100–108.

2. Garaï ME, Slotow R, Reilly B, Carr RD (2004) History and success of elephant re-introductions to small fenced reserves in South Africa. Pachyderm 37: 28–36.

3. Slotow R, Hunter LTB (2008) Reintroduction decisions taken at the incorrect social scale devalue their conservation contribution: African lion in South Africa. In: Hayward MW, Somers MJ, editors. The Reintroduction of Top-order Predators. Oxford, UK: Blackwells Publishing.

4. Hunter L, Skinner JD, Pretorius K, Carlisle LC, Rickleton M, et al. (2007) Reintroducing lions Panthera leo to northern Kwazulu-Natal, South Africa: short-term biological and technical successes but equivocal long-term conservation. Oryx 41: 196–204.

5. Hayward MW, Adendorff J, O'Brien J, Sholto-Douglas A, Bissett C, et al. (2007) Practical considerations for the reintroduction of large, terrestrial, mammalian predators based on reintroductions to South Africa's Eastern Cape Province. The Open Conservation Biology Journal 1: 1–11 doi 10.2174/1874-8392/07 http://www.benthamopen.org/crdsb/?TOCONSBJ/2007/00000001/00000001/1TOCONSBJ.sgm.

6. Gittleman JL, Harvey PH (1982) Carnivore home-range size, metabolic needs and ecology. Behav ecol Sociobio 10: 57–63.

7. MacDonald DW (1983) The ecology of carnivore social behavior. Nature 301: 379–383.

8. Bertram B (1973) Lion population regulation. E Afr Wildl J 2: 215–225.

9. Van Orsdol KG, Hanby JP, Bygott JD (1985) Ecological correlates of lion social organisation (Panthera leo). J Zool Lond 206: 97–112.

10. Funston PJ, Mills MGL, Biggs HC, Richardson PRK (1998) Hunting by male lions: ecological influences and socioecological implications. Anim Behav 56: 1333–1345.

11. Packer C, Scheel D, Pusey AE (1990) Why lions form groups: food is not enough. Am Nat 136: 1–17.

12. Packer C, Gilbert DA, Pusey AE, O'Brien SJ (1991) A molecular genetic analysis of kinship and cooperation in African lions. Nature 351: 562–565.

13. Packer C, Pusey A (1983) Adaptations of female lions to infanticide by incoming males. Am Nat 121: 716–728.

14. Packer C (2000) Inranticide is no fallacy. Am Anth 102: 829–857.

15. Packer C, Pusey A (1997) Divided we fall: cooperation among lions. Scientific American May 32–39.

16. Grinnel J, McComb K (2001) Roaring and social communication in African lions: the limitations imposed by listeners. Anim Behav 62: 93–98.

17. Funston PJ (1999) Predator-prey relationship between lions and large ungulates in the Kruger National Park. PhD Thesis. University of Pretoria.

18. Starfield AM, Furniss PR, Smuts GL (1981) A model of lion population dynamics as a function of social behavior. In: Fowler CW, editor. Dynamics of large mammal populations. John Wiley & Sons. pp. 121–134.

19. Spong G (2002) Space use in lions, Panthera leo, in the Selous Game Reserve: social and ecological factors. Behav Ecol Sociobiol 52: 303–307.

20. Rutherford MC, Westfall RH (2003) Biomes of Southern Africa: An objective categorisation. National Botanical Institute, Pretoria.

21. Low AB, Rebelo AC (1998) Vegetation of South Africa, Lesotho and Swaziland. Department of Environmental Affairs and Tourism, Pretoria.

22. Lehmann MB, Funston PJ, Owen CR, Slotow R (2008) Feeding behavior of lions (Panthera leo) on a small reserve. S Afr J Wildl Res 38(1): 66–78.

23. Mills MGL (1996) Methodological advances in capture, census, and food-habits studies of large African carnivores. In: Gittleman JL, editor. Carnivore Behavior, Ecology and Evolution. London: Comstock Publishing Association. pp. 223–242.

24. Hooge PN (1999) Animal movement analysis ArcView extension. Biological Science Centre USGS-BRD, Alaska.

25. Burt WH (1943) Territoriality and home range concepts as applied to mammals. J Mammal 24: 346–352.

26. Seaman DE, Powell RA (1996) An evaluation of the accuracy of kernel density estimators for home range analysis. Ecology 77: 2075–2085.

27. Mizutani F, Jewell PA (1998) Home-range and movements of leopards on a livestock ranch in Kenya. J Zool Lond 244: 269–286.

28. Apps CD (1999) Space-use, diet demographics and topographic associations of Lynx in the Canadian Rocky Mountains: A study. USDA Forest Service Research Papers. pp. 1–11.

29. Hopcraft JGC, Sinclair ARE, Packer C (2005) Planning for success: Serengeti lions seek prey accessibility rather than abundance. J Anim Ecol 74: 559–566.

30. Schaller GB (1972) The Serengeti Lion: a study of predator-prey relations. Chicago: University of Chicago Press.

31. Hanby JP, Bygott JD (1987) Emigration of subadult lions. Anim Behav 35: 161–169.

32. Stander PE (1991) Demography of lions in the Etosha National Park, Namibia. Madoqua 18: 1–9.

33. Druce D, Genis H, Braak J, Greatwood S, Delsink A, et al. (2004) Population demography and spatial ecology of a reintroduced lion population in the Greater Makalali Conservancy, South Africa. Koedoe 47: 103–118.

34. Borger L, Franconi N, De Michele G, Gantz A, Meschi F, et al. (2006) Effects of sampling regime on the mean and variance of home range size estimates. J Anim Ecol 75: 1393–1405.

35. Whyte J (1985) The present ecological status of the Blue Wildebeest (Connochaetes taurinus taurinus, Burcell, 1923) in the central district of the Kruger National Park. Partial fulfilment of the requirements for the degree of Masters of Science in the Institute of Natural Resources. University of Natal.

36. Funston PJ (1997) Review and progress report on the Northern Plains Project May 1997. Internal Report 5/97, Scientific Services, Skukuza, Kruger National Park.

37. Bradbury JW, Vehrencamp SL (1976) Social organisation and foraging in emballonurid bats. I. Field studies. Behav Ecol Sociobiol 1: 337–381.

38. Grinnel J, McComb K (1996) Maternal grouping as a defence against infanticide by males: evidence from field playback experiments on African lions. Behav Ecol 1: 55–59.

39. Pusey AE, Packer C (1987) The evolution of sex-biased dispersal in lions. Behavior 101: 275–310.

40. Hunter L (1999) The socio-ecology of re-introduced lions in small reserves: comparisons with established populations and the implications for management in enclosed conservation areas. Preliminary Project report. Durban: University of Natal.

41. Funston PJ, Mills MGL, Biggs HC (2001) Factors affecting hunting success of male and female lions in the Kruger National Park. J Zool Lond 253: 419–431.

CITATION

Introduced Mammalian Predators Induce Behavioral Changes in Parental Care in an Endemic New Zealand Bird

Melanie Massaro, Amanda Starling-Windhof, James V. Briskie
and Thomas E. Martin

ABSTRACT

The introduction of predatory mammals to oceanic islands has led to the extinction of many endemic birds. Although introduced predators should favour changes that reduce predation risk in surviving bird species, the ability of island birds to respond to such novel changes remains unstudied. We tested whether novel predation risk imposed by introduced mammalian predators has altered the parental behavior of the endemic New Zealand bellbird (Anthornis melanura). We examined parental behavior of bellbirds at three woodland sites in New Zealand that differed in predation risk: 1) a mainland site with exotic predators present (high predation risk), 2) a mainland site with exotic predators experimentally removed (low risk recently) and, 3) an off-shore island where exotic predators were never introduced (low risk always). We also compared parental behavior of bellbirds with two closely related Tasmanian honeyeaters (Phylidonyris spp.) that evolved with native nest predators (high risk always). Increased nest predation risk has been postulated to favour reduced parental activity, and we tested whether island bellbirds responded to variation in predation risk. We found that females spent more time on the nest per incubating bout with increased risk of predation, a strategy that minimised activity at the nest during incubation. Parental activity during the nestling period, measured as number of feeding visits/hr, also decreased with increasing nest predation risk across sites, and was lowest among the honeyeaters in Tasmania that evolved with native predators. These results demonstrate that some island birds are able to respond to increased risk of predation by novel predators in ways that appear adaptive. We suggest that conservation efforts may be more effective if they take advantage of the ability of island birds to respond to novel predators, especially when the elimination of exotic predators is not possible.

Introduction

The majority of bird extinctions since 1800 have occurred on islands and the main cause of these extinctions has been the introduction of exotic predators [1], [2] often in close association with drastic habitat alterations [3], [4]. The impact of introduced predators on the native avifauna of oceanic islands is particularly profound because the birds evolved largely in the absence of many predators [e.g., 5]. In continental areas, birds and predators have co-evolved over millions of years, and many behavourial and life history traits vary adaptively with risk of predation [6]–[9]. In contrast, native birds on predator-free islands appear to have lost adaptations to avoid terrestrial predators. Instead, they exhibit behaviors and life history traits (e.g. tameness, loss of flight, large size, low fecundity) that predispose them to population crises when predatory animals are introduced [10], [11], suggesting that they are evolutionarily 'trapped' [12], [13]. However, island birds 'trapped' by exotic predators are not necessarily condemned to extinction [13]. The relative risk of extinction will depend on the ability of a species to adjust behavioral traits or evolve in response to exotic predators. Yet, studies of trait changes in response to novel changes in predation risk among island birds are lacking. Here we present a detailed study of responses in island honeyeaters to variation in current and historic predation risk on New Zealand and Tasmania, Australia.

New Zealand provides a typical example of problems arising from introduction of exotic predators. Extinctions of birds on oceanic islands such as New Zealand have been directly linked to human-induced habitat destruction and the introduction of predatory mammals [2]–[4], [14]–[16]. New Zealand was first settled by Maori in ~1300, and then by Europeans beginning in 1769. Both settlement phases were associated with the introduction of exotic mammalian predators; Maori introduced the Polynesian rat (Rattus exulans), while Europeans introduced the house mouse (Mus domesticus), two additional species of rats (R. rattus and R. norvegicus), the hedgehog (Erinaceus europaeus), the domestic cat (Felis catus), three mustelids (Mustela erminea, M. furo and M. nivalis) and the brushtail possum (Trichosurus vulpecula). These introductions contributed to the extinction of ~40% of all non-marine bird species in New Zealand [3], [16], [17] and pose a major threat to the survival of the remaining avifauna [18].

In this study we tested whether an endemic New Zealand songbird, the bellbird (Anthornis melanura), altered its parental behavior and life history traits in ways that might adapt it to the novel predation risk from introduced mammalian predators. In particular, increased parental activity at a nest can attract predators and increase nest predation rates [19]. Bird species adaptively differ in their rates of parental feeding visits to the nest during the nestling period related to risk of predation [8], [9], [20], [21]. Birds can also achieve lower activity during incubation by reducing the number of visits per unit time and increasing the length of time per bout sitting on the nest [22], [23]. Increased nest predation risk could also favour the evolution of shorter incubation periods to minimise the total time a nest is at risk [24], [25]. Bellbirds, therefore, might respond adaptively to the increased risk of predation by novel predators by lowering parental activity, increasing length of incubation on-bouts, and shortening incubation periods.

We examined behavioral and life history responses in the bellbird to changing nest predation risk on differing time scales. A few New Zealand offshore islands have never had exotic predators and provide a benchmark to compare with populations on the main New Zealand islands that have been exposed to novel predators beginning 700 years ago. As small islands might alter life history traits independently of predation risk (e.g. due to higher density), we also conducted a predator removal experiment on the mainland New Zealand to further examine whether bellbirds assess current predation risk and alter their behavior and life history traits accordingly. Finally, to examine the effects of historical predation risk, we compared parental behavior of bellbirds in New Zealand with two related honeyeaters (Phylidonyris spp.) in Tasmania, Australia. Tasmanian honeyeaters evolved with a variety of native mammalian predators, yet share a common ancestor with the bellbird. Thus, honeyeaters in Tasmania and New Zealand provide an opportunity to examine differences in responses to differing long-term exposure to mammalian predators.

Methods

Study Areas

The bellbird is a medium-sized honeyeater (26–34 g) endemic to New Zealand [26]. The abundance and range of bellbirds has decreased since human settlement, but they survived within most native forest areas on the main islands of New Zealand and on several offshore islands [26]. Bellbirds were studied in three locations: (1) on Aorangi Island (35°28′ S, 174°44′ E), a forested island of approximately 66 ha, ~22 km off the east coast of the North Island where exotic mammalian predators have never existed, (2) in Waiman Bush (42°20′ S, 173°40′ E), a 65 ha native forest located 15 km from Kaikoura where all exotic predators were continuously removed throughout the year from 2004 to 2007, and (3) in Kowhai Bush (42°22′ S, 173°36′ E), a 240 ha native forest located 10 km from Kaikoura, South Island where all exotic predators are present.

Parts of Aorangi Island were cultivated by Maori until 1820, but it was then abandoned and the island has remained uninhabited and declared a nature reserve in 1929. Polynesian rats were never introduced during Maori settlement on the island and to this day it remains free of all introduced predatory mammals. The island is far enough from the mainland that gene flow is likely to have been minimal, a possibility that is supported by the slightly different coloration of birds compared to the mainland [27]. The only potential predators present on the island are native and include Australasian harrier (Circus approximans), long-tailed cuckoo (Eudynamys taitensis), shining cuckoo (Chrysococcyx lucidus) and perhaps the large Duvaucel's gecko (Hoplodactylus duvaucelii). In contrast, the mainland site at Kowhai Bush includes all species of exotic predators plus native avian predators. The birds in Kowhai Bush have co-existed with exotic mammalian predators for at least the last 700 years, a situation typical of that faced by all surviving native species on the main islands of New Zealand. Waiman Bush is at the same elevation and includes the same native forest habitat and avifauna as Kowhai Bush. The two sites are separated by about 5 km of mostly cleared agricultural land although connected by continuous forest at a higher elevation. Beginning in 2004, all species of exotic mammalian predators were removed from Waiman Bush using 38 tunnel traps to control mustelids, rats and hedgehogs, 8 Timms traps for possum and cat control, and 52 poison bait stations controlling rats and possums. A total of 90 stoats, 24 ferrets, 24 weasels, 23 possums, 137 rats, 218 hedgehogs, and 32 cats were caught in traps during this period and an additional unknown number of animals were killed by poison from the bait stations (or through secondary poisoning). It is not possible to permanently remove all predators from mainland sites but similar efforts to control predators at other New Zealand sites have lead to increased nest success and population increases of many native birds [28], [29], a general pattern that is evident on our study site as well [30].

While Aorangi Island is located further north than Kowhai and Waiman Bush (7° on a north south axis), all three sites are lowland coastal forests with a similar canopy structure and experience a similar maritime climate. The composition of the forest differs slightly between Aorangi and the two mainland sites and this is reflected in the nest sites selected by bellbirds. On Aorangi Island most bellbird nests are built in weeping matipo (Myrsine divaricata), the native vine Muehlenbeckia spp., and Coprosma macrocarpa [for details on vegetation see 31], [32], while in Kowhai and Waiman Bush bellbirds generally nest in kanuka (Leptospermum ericiodes) and in the shrub Coprosoma robusta. Nests at all three sites are placed so that they are well-concealed by dense vegetation and it is unlikely that any differences in parental behavior between sites was due to differences in nest site placement.

We also studied honeyeaters in Tasmania, Australia to examine traits in a site where native predators have always existed. It is unknown which Australian honeyeater is phylogenetically the closest relative to the New Zealand bellbird [33], but we selected two species of native honeyeaters (crescent honeyeater Phylidonyris pyrrhoptera and New Holland honeyeater P. novaehollandiae) in Tasmania that are in the same family as bellbirds and so have a common ancestor. The crescent and New Holland honeyeaters are of similar size and morphology as the New Zealand bellbird, and the habitat preferences, life history traits, mating systems and parental behaviors are also similar among these three species. For example, mean clutch size is 2.8, 2.2 and 3.1 eggs in the crescent honeyeater, New Holland honeyeater and the bellbird, respectively; in all three species both the incubation and nestling period are 13–14 days long; all three species are socially monogamous and share parental care, whereby females are solely responsible for nest construction and incubation, but both sexes feed nestlings [26], [34]–[37]. The two Australian honeyeaters were studied in the Scamander Forest Reserve near St. Helens, Tasmania (41°27' S, 148°15' E). This is a 100 ha native forest block that is not subject to logging or hunting and contains a wide range of native predatory mammals and snakes. We chose to study honeyeaters in Tasmania because it is located at a similar latitude and experiences a similar maritime climate to our New Zealand study sites. Scamander Forest Reserve is also located at the same elevation as our New Zealand study sites and the forest structure is similar although it is dominated by gums (Eucalyptus spp.) with an acacia (Acacia spp.) and fern understorey. Like the bellbird, the open-cup nests of the Tasmanian honeyeaters were placed in the shrub and canopy layer and well-concealed by surrounding vegetation.

Data on life history and nesting behaviors were collected at all study sites from October until December each year (Aorangi: 2004–2005; Waiman Bush: 2004– 2006; Kowhai Bush: 1998–2007; Tasmania: 2004–2005). Nests were found by following adults, and nests were monitored every 2–5 days to record nest success. Daily nest predation rates were calculated using the Mayfield method [38], [39]. We followed Hensler & Nichols [40] to calculate standard errors for Mayfield's daily predation probabilities, and analysed differences in daily predation rates using the CONTRAST program [41]. Clutch size was determined for accessible nests. For nests found during nest-building or egg laying, we measured incubation periods as the period from last egg laid to the last egg hatched (to an accuracy of 2 days or less). To estimate parental visitation rates, we videotaped nests during both the incubation and the nestling stage using portable Sony Hi8 video cameras. Nests were filmed for the first 6 hours of the day, starting within 30 min of sunrise, except on Aorangi in 2004 when nests were taped later in the morning. Despite this difference in protocol, we pooled data across years because we did not detect significant differences on Aorangi between 2004 and 2005 in parental visits during incubation ($F_{1,17}$ = 0.001, p = 0.98), incubation attentiveness ($F_{1,17}$ = 3.6, p = 0.08), mean on-bout length ($F_{1,16}$ = 0.14, p = 0.72), mean off-bout length ($F_{1,16}$ = 3.05, p = 0.1), and nestling feeding visits ($F_{1,20}$ = 2.86, p = 0.11) . Nests were filmed throughout the incubation period although we avoided filming nests within the first few days after laying. Incubation videos were scored for number of parental visits to the nest, nest attentiveness (measured as percentage of 6 h that females sat on the nest), and mean duration of incubation on and off-bouts [25], [42]. One incubation video from Aorangi was excluded because the female was extensively fed by her mate while on the nest, the only example of this behavior we noted. As we expected higher visitation rates on the island with no exotic predators, the high rate of male visits to this nest increased our estimate of total visitation rate, such that the removal of this outlier makes our test of higher visitation rates in the absence of exotic predators more conservative. Nests with nestlings were videotaped within one day of nestlings breaking primary pinfeathers, to control for differences in developmental rates between locations or species and to quantify rates of parental visits to the nest to feed nestlings [9], [21].

Locating bellbird and honeyeater nests is time consuming, and given the high probability of nest failure (often before data on parental investment could be collected), multiple seasons across sites had to be sampled to increase sample sizes. To control for repeat sampling of females or pairs across seasons (no repeat sampling of females or pairs occurred within a season), we individually color-banded

77 and 64 adult bellbirds on Aorangi Island in 2004 and in Kowhai Bush from 2000–2006, respectively. On Aorangi Island, nests of only one female were found in both seasons, and in Kowhai Bush nests of two females were found in two consecutive seasons. In all three cases banded females were paired with an unbanded male of unknown breeding history. To avoid repeat sampling of females, we randomly selected one incubation video and one nestling stage video per female for the analysis. The wide spatial distribution of filmed nests at all sites also minimised the chances of resampling unbanded birds more than once.

We tested whether life history traits and parental behaviors varied among locations by conducting ANOVAs after ensuring that the assumptions of an ANOVA were met (homoscedasticity and normality). We used LSD post-hoc tests to examine differences among individual sites when the ANOVA was significant. All means are reported±standard error.

Results

Daily nest predation rates for bellbirds were significantly lower on Aorangi Island (no exotic predators present) than in Waiman Bush (exotic predators removed) and Kowhai Bush (exotic predators present) (Figure 1). Daily predation rate was lower in Waiman Bush than in Kowhai Bush although the difference was not significant (Figure 1). Over a nesting cycle spanning about 30 days, these differences give a probability of a nest surviving to fledge of 65% for bellbirds on Aorangi, 39% for bellbirds in Waiman Bush and only 29% for bellbirds in Kowhai Bush.

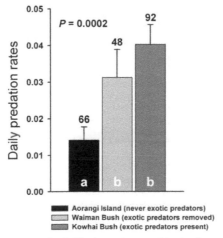

Figure 1. Daily nest predation rates for bellbirds on Aorangi Island, where exotic predators were never introduced, in Waiman Bush, where exotic predators were removed, and in Kowhai Bush, where all exotic predators are present. Shared letters within bars denote non-significant (i.e., p>0.05) statistical differences using the CONTRAST program [41]. Figures above bars are the number of nests in each study site.

Parental behavior varied among bellbird populations with the varying levels of predation risk. The number of parental visits to the nest during incubation varied significantly among the four sites ($F_{3,61}$ = 5.12, p = 0.003, Figure 2). Bellbirds on Aorangi Island, where exotic predators were never introduced, and in Waiman Bush where predators were removed, visited their nests at similar rates. At both these sites bellbirds visited their nests more frequently than at Kowhai Bush, where a variety of exotic predators are present, and honeyeaters in Tasmania that evolved with native predators (Figure 2). We obtained the same results when we controlled for day of incubation and only included nests that were filmed during the middle of incubation (days 4–8 of a 14 day incubation period); parental visitation rates during the middle of incubation differed among sites ($F_{3,27}$ = 5.59, p = 0.004) in a similar pattern: the rate on Aorangi was similar to the rate in Waiman Bush, which was higher than the visit rate in Kowhai Bush, which in turn was similar to that of honeyeaters in Tasmania.

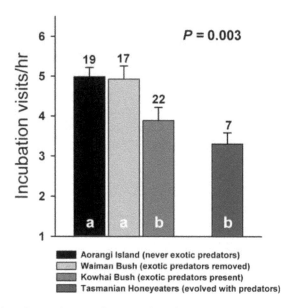

Figure 2. The number of parental visits to the nest per hour during incubation for bellbirds on Aorangi Island, where exotic predators were never introduced, for bellbirds in Waiman Bush, where exotic predators were removed, for bellbirds in Kowhai Bush, where all exotic predators are present, and for honeyeaters in Tasmania, which evolved with a range of native mammalian predators. Shared letters within bars denote non-significant (i.e., p>0.05) statistical differences based on LSD tests. Figures above bars are the number of nests filmed in each study site.

Nest attentiveness (percent time females spend on the nest), in contrast to visit rates, did not differ among sites ($F_{3,61}$ = 0.346, p = 0.8). Mean nest attentiveness by females was 67.8% (±1.59, n = 19) on Aorangi , 66.2% (±1.91, n = 17) in

Waiman Bush, 68.5% (±1.73, n = 22) in Kowhai Bush, and 68.6% (±2.31, n = 7) in Tasmania. Thus bellbirds did not alter the total time spent incubating with increased predation risk but instead decreased the number of visits made to and from the nest by changing bout lengths.

The mean time females spent on the nest per incubating bout differed among the four sites ($F_{3,60}$ = 3.58, p = 0.019, Figure 3a). Duration of on-bouts during incubation were similar on Aorangi (no exotic predators present) and Waiman Bush (predators controlled), but both were shorter than in Kowhai Bush (no predator control; Figure 3a). Honeyeaters in Tasmania had similar durations of on-bouts as bellbirds in Kowhai Bush and Waiman Bush, but significantly longer than bellbirds on Aorangi (Figure 3a). Duration of off-bouts (time females spend away from the nest during each recess) also differed among sites ($F_{3,60}$ = 5.75, p = 0.002) and mirrored the pattern observed in on-bouts with bellbirds showing increasing times spent away from the nest as predation risk increased: bellbirds had the shortest off-bouts on Aorangi and the longest in Kowhai Bush (Figure 3b). Honeyeaters in Tasmania had off-bouts similar to that observed in bellbirds in Kowhai Bush (Figure 3b).

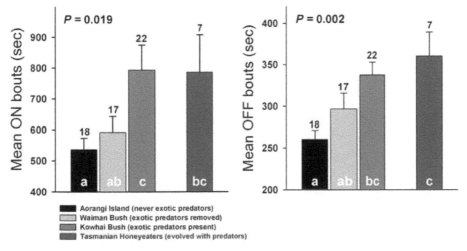

Figure 3. The mean time females spent on the nest per incubating bout (mean on-bouts) and the mean time females spent away from the nest (mean off-bouts) for bellbirds on Aorangi Island, where exotic predators were never introduced, for bellbirds in Waiman Bush, where exotic predators were removed, for bellbirds in Kowhai Bush, where all exotic predators are present, and for honeyeaters in Tasmania, which evolved with a range of native mammalian predators. Shared letters within bars denote non-significant (i.e., p>0.05) statistical differences based on LSD tests. Figures above bars are the number of nests filmed in each study site.

Rate of parental feeding visits to the nest during the nestling period differed among all four sites ($F_{3,37}$ = 19.274, p<0.0001) with bellbirds on Aorangi (no predators present) visiting nests more frequently than at either Waiman Bush (predators removed) or Kowhai Bush (predators present; Figure 4). There was no significant difference between Waiman and Kowhai Bush, so our predator removal experiment did not change the feeding behavior of bellbirds. Parental feeding rates at both Waiman and Kowhai Bush were significantly higher than that observed in honeyeaters in Tasmania (Figure 4). When we controlled for brood size by only including nestling videos of 3-chick broods, nestling feeding rates still differed strongly among the sites ($F_{3,19}$ = 9.623, p<0.0001), with the same significant differences among study sites.

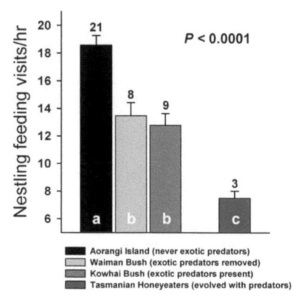

Figure 4. The number of parental visits to the nest per hour to feed nestlings for bellbirds on Aorangi Island, where exotic predators were never introduced, for bellbirds in Waiman Bush, where exotic predators were removed, for bellbirds in Kowhai Bush, where all exotic predators are present, and for honeyeaters in Tasmania, which evolved with a range of native mammalian predators. Shared letters within bars denote non-significant (i.e., p>0.05) statistical differences based on LSD tests. Figures above bars are the number of nests filmed in each study site.

In contrast to parental behavior, bellbirds exhibited little change in other life history traits. Bellbird clutch size did not differ among the three sites in New Zealand ($F_{2,117}$ = 0.251, p = 0.8). Mean clutch sizes were 3.14 eggs (±0.08, n = 57) on Aorangi (exotic predators never introduced), 3.08 eggs (±0.15, n = 12) in Waiman Bush (exotic predators removed experimentally), and 3.06 eggs (±0.09, n = 51) in Kowhai Bush (exotic predators present). The Tasmanian honeyeaters were not included in this analysis as sample sizes were small but all crescent honeyeaters laid 3 eggs (n = 6 nests) and all New Holland honeyeaters laid 2 eggs (n = 3).

Incubation period also did not differ among the three sites in New Zealand ($F_{2,22}$ = 1.073, p = 0.4). Mean incubation periods were 14.1 days (±0.21, n = 11) on Aorangi, 14.6 days (±0.43, n = 7) in Waiman Bush, and 14.9 days (±0.60, n = 7) in Kowhai Bush. We were unable to gather enough information on incubation periods in Tasmanian honeyeaters, but published records indicate both crescent honeyeater (13.2±0.20 days, n = 5) [43] and New Holland honeyeater (13.4±0.12 days, n = 19) [43] have shorter incubation periods than the New Zealand bellbird.

Discussion

We examined how an endemic island bird that evolved largely without terrestrial predators responds to the novel, and an unusually high, risk of predation due to the introduction of multiple, exotic, mammalian predators to New Zealand. We found that the presence of introduced mammalian predators in New Zealand over the past 700 years have induced shifts in parental behavior in the endemic bellbird that appear to be adaptive. These changes converge on behaviors seen in other species of honeyeaters endemic to Australia, which co-evolved with a variety of predators, and which presumably evolved to minimise the risk from predation. Our results suggest that bellbirds, and perhaps other endemic island birds, are not stuck in an evolutionary "trap" as has been proposed, but instead have some capacity to adapt to novel changes in environment including that posed by the introduction of exotic predators.

Following a hypothesis by Skutch [19], parental activity at the nest can attract the attention of predators and increase nest predation risk [8], [9], [20]. An adaptive response would be to reduce activity with increasing predation risk [9], [23], [44], [45]. Our results suggest such adaptive responses in bellbirds: bellbirds on an offshore island without exotic predators and on the mainland predator-removal site had shorter on- and off-bouts that yielded higher parental visit rates than birds that have now co-existed with exotic predators for c. 700 years (Figures 1, 2). Moreover, bellbirds that coexist with exotic predators had long on- and off-bouts which reduced visit rates to a level similar to their Tasmanian relatives that evolved with mammalian predators (Figures 1, 2). Thus, our results suggest that New Zealand bellbirds are able to respond to exotic nest predation risk by altering their incubation behavior in a manner similar to related species of honeyeater in Tasmania in a period not exceeding 700 years.

In contrast, historical differences appear to remain for parental activity during the nestling period. Bellbirds that co-exist with exotic predators decreased their nestling feeding rates compared to birds on the offshore island without predators, as expected under an adaptive shift, but did not increase feeding when we experimentally removed predators. Nevertheless, bellbirds on the mainland still fed their nestlings twice as often as honeyeaters in Tasmania suggesting the persistence of higher rates of nest visitation in the presence of exotic predators. Although we cannot rule out differences in diet between these species as a potential explanation, our results are also consistent with the view that bellbirds on the New Zealand mainland appear to be adapting to exotic predators but they retain behavioral traits present in naive populations.

Some life history traits appear to show little response to predation risk. Increased nest predation risk has been argued to favour decreased clutch size to reduce the number of nest visits that could attract the attention of predators and the overall period when the nest is vulnerable to nest predators [6], [19], [46]. Presentations of predator models have yielded clutch size reductions [47], [48], while a predator-removal experiment yielded no change in clutch size among eight coexisting passerine species [21]. Similar to the latter result, bellbirds did not reduce their clutch size and laid an average of 3 eggs (range of 2–4 eggs) at all sites despite the difference in predation risk among sites. Thus, this trait was less responsive than parental visitation rates to predation risk.

Female bellbirds increased the length of on- and off-bouts and thereby reduced parental activity, but did not increase their total incubation attentiveness with increased predation risk. Conway & Martin [23] found the same pattern across North American passerine species. Increased attentiveness can potentially shorten incubation and reduce nest predation risk [24], [25], but the lack of change in nest attentiveness is consistent with the lack of change in incubation periods that we observed. Increased attentiveness might compromise adult survival [24], [49]. Female bellbirds incubate alone, and males generally do not feed females during incubation (apart from the one exception noted in the Methods). As a result, females might require a set amount of time off the nest to replenish their resources to minimize mortality costs to themselves. Moreover even if females increased attentiveness, offspring might not have the physiological means to accelerate embryonic development. New Zealand songbirds typically have very long developmental periods that might reflect intrinsic mechanisms to enhance offspring quality and longevity that may not be altered by attentiveness. Nonetheless, the end result is that nest attentiveness and incubation periods did not change with nest predation risk and further demonstrates that traits differ in their responsiveness to nest predation risk.

Island species are often thought to lack the ability to adapt to novel predators, but our data suggest that at least some traits have shifted or are plastic in response to predation risk in ways that appear adaptive. Adaptive phenotypic evolution has recently received considerable attention in the literature, because it offers opportunities for new, innovative approaches to ecosystem management and conservation efforts [13], [50]–[53]. An eco-evolutionary perspective has been promoted [13], [50], [53], whereby contemporary evolution arising from the novel interaction between invasive and native species is considered. In practice, a study aimed at detecting the minimum thresholds of management required to induce the responses that allow the long-term persistence of native bird populations is now necessary to develop such a new management tool further. To this effect it would be useful to replicate our study using further removal experiments that varied in the extent to which predation risk is decreased, and to examine whether other island species have responded in a similar fashion as we found with bellbirds.

One of the main problems when attempting to measure contemporary evolution in native island birds in response to the introduction of exotic predators is that few island bird populations still exist in habitats that have not been affected by human-mediated changes. While changes in morphology and genetic variation can by studied by comparing current populations with museum collections [54], [55], the measurement of behavioral responses to introduced predators requires live bird populations in areas that remain relatively undisturbed by anthropogenic effects. Here, we had the opportunity to study such behavioral responses because of the unique situation in New Zealand where exotic predators were introduced, but a few offshore islands remained undisturbed. While we report a phenotypic change in parental behaviors of bellbirds, we are uncertain about the relative contributions of genetic and non-genetic effects. Regardless, we believe that an improved understanding of the adaptive potential of species facing drastic environmental change and the rate at which such threatened species can achieve phenotypic adaptation can aid future management efforts for the conservation of threatened island bird species.

Acknowledgements

This project would not have been possible without the dedicated and enthusiastic efforts of Barry and Jenny Dunnett to control introduced predators in Waiman Bush over the last 4 years. We thank Lainie Berry, Lucy Birkinshaw, Anna Hosking, James Muir, Danielle Shanahan, Lisa Shorey, Jarom Stanaway and Sabrina Taylor for help in the field. This study was approved by Animal Ethics Committee of the University of Canterbury, the New Zealand Department of Conservation, and the Tasmanian Department of Primary Industries, Water and Environment. We thank Peter Gaze, Keith Hawkins, Richard Parrish and Monica Valdes from the Department of Conservation for providing assistance and help with logistics. We are grateful to Jeroen Jongejans, Kate Malcolm and the team of Dive!Tutukaka who provided a life-line to the mainland during field work on Aorangi Island. Jack van Berkel provided assistance during our many stays at the field station in Kaikoura. Katie Dugger provided statistical advice and Ben Sheldon and two anonymous reviewers made helpful comments to an earlier draft.

Authors' Contributions

Conceived and designed the experiments: JB MM. Performed the experiments: TM JB MM AS. Analyzed the data: MM. Contributed reagents/materials/analysis tools: TM JB MM. Wrote the paper: JB MM. Other: Made valuable comments to an earlier draft of the manuscript: TM. Made comments to an earlier draft of the manuscript: AS.

References

1. Stattersfield AJ, Capper DR, Dutson GCL (2000) Threatened Birds of the World: the Official Source for Birds on the IUCN Red List. Cambridge: BirdLife International.

2. Blackburn TM, Cassey P, Duncan RP, Evans KL, Gaston KJ (2004) Avian extinction and mammalian introductions on oceanic islands. Science 305: 1955–1958.

3. Didham RK, Ewers RM, Gemmell NJ (2005) Comment on "Avian extinction and mammalian introductions on oceanic islands." Science 307: 1412a.

4. Didham RK, Tylianakis JM, Gemmell NJ, Rand TA, Ewers RM (2007) Interactive effects of habitat modification and species invasion on native species decline. Trends Ecol Evol 22: 489–496.

5. Savidge JA (1987) Extinction of an island forest avifauna by an introduced snake. Ecology 68: 660–668.

6. Martin TE (1995) Avian life history evolution in relation to nest sites, nest predation and food. Ecol Monogr 65: 101–127.

7. Briskie JV, Martin PR, Martin TE (1999) Nest predation and the evolution of nestling begging calls. Proc R Soc B 266: 2153–2159.

8. Martin TE, Scott J, Menge C (2000) Nest predation increases with parental activity: separating nest site and parental activity effects. Proc R Soc B 267: 2287–2293.

9. Martin TE, Martin PR, Olson CR, Heidinger BJ, Fontaine JJ (2000) Parental care and clutch sizes in North and South American Birds. Science 287: 1482–1485.

10. Owens IPF, Bennett PM (2000) Ecological basis of extinction risk in birds: habtitat loss versus human persecution and introduced predators. Proc Natl Acad Sci USA 97: 12144–12148.

11. Duncan RP, Blackburn TM (2004) Extinction and endemism in the New Zealand avifauna. Global Ecol Biogeogr 13: 509–517.

12. Schlaepfer MA, Runge MC, Sherman PW (2002) Ecological and evolutionary traps. Trends Ecol Evol 17: 474–480.

13. Schlaepfer MA, Sherman PW, Blossey B, Runge MC (2005) Introduced species as evolutionary traps. Ecol Letters 8: 241–246.

14. Olson SL, James HF (1982) Fossil birds from the Hawaiian Islands: evidence for wholesale extinction by man before western contact. Science 217: 633–635.

15. Blackburn TM, Petchey OL, Cassey P, Gaston KJ (2005) Functional diversity of mammalian predators and extinction in island birds. Ecology 86: 2916–2923.

16. Diamond JM, Veitch CR (1981) Extinctions and introductions in the New Zealand avifauna: cause and effect? Science 211: 499–501.

17. Holdaway RN (1999) Introduced predators and avifaunal extinction in New Zealand. In: McPhee RD, editor. Extinctions in near time. New York: Kluwer Academic/Plenum. pp. 189–238.

18. Worthy TH, Holdaway RN (2002) The lost world of the moa. Bloomington: Indiana University Press.

19. Skutch AF (1949) Do tropical birds rear as many young as they can nourish? Ibis 91: 430–455.

20. Muchai M, du Plessis MA (2005) Nest predation of grassland bird species increases with parental activity at the nest. J Avian Biol 36: 110–116.

21. Fontaine JJ, Martin TE (2006) Parent birds assess nest predation risk and adjust their reproductive strategies. Ecol Letters 9: 428–434.

22. Weathers WW, Sullivan KA (1989) Nest attentiveness and egg temperature in the Yellow-eyed Junco. Condor 91: 628–633.

23. Conway CJ, Martin TE (2000) Evolution of passerine incubation behavior: influence of food, temperature, and nest predation. Evolution 54: 670–685.

24. Martin TE (2002) A new view for avian life history evolution tested on an incubation paradox. Proc R Soc B 269: 309–316.

25. Martin TE, Auer SK, Bassar RD, Niklison AM, Lloyd P (2007) Geographic variation in avian incubation periods and parental influences on embryonic temperature. Evolution 61-11: 2558–2569.

26. Heather BD, Robertson HA (2005) The field guide to the birds of New Zealand. Revised edition. Auckland: Penguin Books.

27. Bartle JA, Sagar PM (1987) Intraspecific variation in the New Zealand bellbird Anthornis melanura. Notornis 34: 253–306.

28. O'Donnell CFJ, Dilks PJ, Elliott GP (1996) Control of a stoat (Mustela erminea) population irruption to enhance mohua (yellowhead) (Mohoua ochrocephala) breeding success in New Zealand. New Zealand J Zool 23: 279–286.

29. Saunders A, Norton DA (2001) Ecological restoration at mainland islands in New Zealand. Biol Cons 99: 109–119.

30. Starling A (2006) Behavioral plasticity of life history traits in the New Zealand avifauna. Unpubl MSc Thesis, University of Canterbury.

31. Sagar PM (1985) Breeding of the bellbird on the Poor Knights Islands, New Zealand. New Zeal J Zool 12: 643–648.

32. de Lange PJ, Cameron EK (1999) The vascular flora of Aorangi Island, Poor Knights Islands, northern New Zealand. New Zeal J Botany 37: 433–468.

33. Driskell AC, Christidis L (2004) Phylogeny and evolution of the Australo-Papuan honeyeaters (Passeriformes, Meliphagidae). Mol Phylogen Evol 31: 943–960.

34. Recher HF (1977) Ecology of co-existing white-cheeked and New Holland Honeyeaters. Emu 77: 136–142.

35. McFarland DC (1986) Breeding behavior of the New Holland honeyeater Phylidonyris novaehollandiae. Emu 86: 161–167.

36. Clarke RH, Clarke MF (1999) The social organization of a sexually dimorphic honeyeater - the crescent honeyeater Phylidonyris pyrrhoptera, at Wilsons Promontory, Victoria. Austral J Ecol 204: 604–654.

37. Clarke RH, Clarke MF (2000) The breeding biology of the crescent honeyeater Phylidonyris pyrrhoptera, at Wilsons Promontory, Victoria. Emu 100: 115–124.

38. Mayfield HF (1961) Nesting success calculated from exposure. Wilson Bull 73: 255–261.

39. Mayfield HF (1975) Suggestions for calculating nest success. Wilson Bull 87: 456–466.

40. Hensler GL, Nichols JD (1981) The Mayfield method of estimating nesting success: a model, estimators and simulation results. Wilson Bull 93: 42–53.

41. Sauer JR, Williams BK (1989) Generalized procedures for testing hypotheses about survival or recovery rates. J Wildl Manage 53: 137–142.

42. Martin TE, Ghalambor CK (1999) Males feeding females during incubation. I. Required by microclimate or constrained by nest predation? Am Nat 153: 131–139.

43. Higgins PJ, Peter JM, Steele WK, editors. (2001) Handbook of Australian, New Zealand and Antarctic birds. Vol. 5. Tyrant flycatchers to chats. South Melbourne, Australia: Oxford University Press.

44. Eggers S, Griesser M, Ekman J (2004) Predator-induced plasticity in nest visitation rates in the Siberian Jay (Perisoreus infaustus). Behav Ecol 16: 309–315.

45. Ferretti V, Llambias PE, Martin TE (2005) Life-history variation of a neotropical thrush challenges food limitation theory. Proc R Soc B 272: 769–773.

46. Roff DA (1992) The evolution of life histories: theory and analysis. New York: Chapman and Hall.

47. Doligez B, Clobert J (2003) Clutch size reduction as a response to increased nest predation rate in the Collard Flycatcher. Ecology 84: 2582–2588.

48. Eggers S, Griesser M, Nystrand M, Ekman J (2006) Predation risk induces changes in nest-site selection and clutch size in the Siberian jay. Proc R Soc B 273: 701–706.

49. Visser ME, Lessells CM (2001) The costs of egg production and incubation in great tits (Parus major). Proc R Soc B 268: 1271–1277.

50. Ashley MV, Willson MF, Pergams ORW, O'Dowd DJ, Gende SM, et al. (2003) Evolutionarily enlightened management. Biol Cons 111: 115–123.

51. Stockwell CA, Hendry AP, Kinnison MT (2003) Contemporary evolution meets conservation biology. Trends Ecol Evol 18: 94–101.

52. Carroll SP, Hendry AP, Reznick DN, Fox CW (2007) Evolution on ecological time-scales. Funct Ecol 21: 387–393.

53. Kinnison MT, Hairston NG (2007) Eco-evolutionary conservation biology: contemporary evolution and the dynamics of persistence. Funct Ecol 21: 444–454.

54. Smith TB, Freed LA, Lepson JK, Carothers JH (1995) Evolutionary consequences of extinctions in populations of a Hawaiian honeycreeper. Cons Biol 9: 107–113.

55. Taylor SS, Jamieson IG, Wallis GP (2007) Historic and contemporary levels of genetic variation in two New Zealand passerines with different histories of decline. J Evol Biol 20: 2035–2047.

CITATION

Massaro M, Starling-Windhof A, Briskie JV, and Martin TE. Introduced Mammalian Predators Induce Behavioral Changes in Parental Care in an Endemic New Zealand Bird. PLoS ONE 3(6): e2331. doi:10.1371/journal.pone.0002331. Originally published under the Creative Commons Attribution License, http://creativecommons.org/licenses/by/3.0/

Copyrights

Index